W9-ANV-611

The Symmetric Group
Representations, Combinatorial Algorithms, and
Symmetric Functions

Series Editors

M. Adams, V. Guillemin, *Measure Theory and Probability*

W. Beckner, A. Calderón, R. Fefferman, P. Jones, *Conference on Harmonic
Analysis in Honor of Antoni Zygmund*

G. Chartrand, L. Lesniak, *Graphs & Digraphs, Second Edition*

W. Derrick, *Complex Analysis and Applications, Second Edition*

J. Dieudonné, *History of Algebraic Geometry*

R. Dudley, *Real Analysis and Probability*

R. Durrett, *Brownian Motion and Martingales in Analysis*

R. Epstein, W. Carnielli, *Computability: Computable Functions, Logic,
and the Foundations of Mathematics*

S. Fisher, *Complex Variables, Second Edition*

A. Garsia, *Topics in Almost Everywhere Convergence*

P. Garrett, *Holomorphic Hilbert Modular Forms*

R. Gunning, *Introduction to Holomorphic Functions of Several Variables*
Volume I: Function Theory
Volume II: Local Theory
Volume III: Homological Theory

H. Helson, *Harmonic Analysis*

J. Kevorkian, *Partial Differential Equations: Analytical Solution Techniques*

R. McKenzie, G. McNulty, W. Taylor, *Algebras, Lattices, Varieties,
Volume I*

E. Mendelson, *Introduction to Mathematical Logic, Third Edition*

D. Passman, *A Course in Ring Theory*

B. Sagan, *The Symmetric Group: Representations, Combinatorial
Algorithms, and Symmetric Functions*

R. Salem, *Algebraic Numbers and Fourier Analysis* and L. Carleson,
Selected Problems on Exceptional Sets

R. Stanley, *Enumerative Combinatorics, Volume I*

J. Strikwerda, *Finite Difference Schemes and Partial Differential
Equations*

K. Stromberg, *An Introduction to Classical Real Analysis*

The Symmetric Group:

Representations, Combinatorial Algorithms, and Symmetric Functions

Bruce E. Sagan
Michigan State University

Wadsworth & Brooks/Cole Advanced Books & Software
Pacific Grove, California

37

Brooks/Cole Publishing Company
A Division of Wadsworth, Inc.

Printed in the United States of America

10   9   8   7   6   5   4   3   2   1

Library of Congress Cataloging-in-Publication Data

Sagan, Bruce Eli.
    The symmetric group: representations, combinatorial algorithms, and symmetric functions/Bruce E. Sagan.
        p.   cm. – (Wadsworth & Brooks/Cole mathematics series)
    Includes bibliographical references (p.   ) and index.
    ISBN 0-534-15540-5
    1. Representations of groups.   2. Symmetric functions.   I. Title.   II. Series.
QA171.S24   1991   90-23333
512′.2–dc20   CIP

$$QA$$
$$171$$
$$.S24$$
$$1991$$

Sponsoring Editor: *John Kimmel*
Editorial Assistant: *Jennifer Kehr*
Production Coordinator: *Marlene Thom*
Manuscript Editor: *Linda Thompson*
Cover Design: *Sharon Kinghan*
Printing and Binding: *Arcata Graphics/Fairfield*

Dedication:
2π

Contents

Preface

In recent years there has been a resurgence of interest in representations of symmetric groups (as well as other Coxeter groups). This topic can be approached from three directions: by applying results from the general theory of group representations, by employing combinatorial techniques, or by using symmetric functions. The fact that this area is the confluence of several strains of mathematics makes it an exciting one in which to study and work. By the same token, it is more difficult to master.

The purpose of this monograph is to bring together, for the first time under one cover, many of the important results in this field. To make the work accessible to the widest possible audience, a minimal amount of prior knowledge is assumed. The only prerequisites are a familiarity with elementary group theory and linear algebra. All other results about representations, combinatorics, and symmetric functions are developed as they are needed. Hence this book could be read by a graduate student or even a very bright undergraduate. For researchers I have also included topics from recent journal articles and even material that has not yet been published.

Chapter 1 is an introduction to group representations, with special emphasis on the methods of use when working with the symmetric groups. Because of space limitations, many important topics that are not germane to the rest of the development are not covered. These subjects can be found in any of the standard texts on representation theory.

In Chapter 2, the results from the previous chapter are applied to the symmetric group itself and more highly specialized machinery is developed to handle this case. I have chosen to take the elegant approach afforded by the Specht modules rather than working with idempotents in the group algebra.

The third chapter focuses on combinatorics. It starts with the two famous formulae for the dimensions of the Specht modules: the Frame-Robinson-Thrall hook formula and the Frobenius-Young determinantal formula. The centerpiece is the Robinson-Schensted-Knuth algorithm, which allows us to describe some of the earlier theorems in purely combinatorial terms. A thorough discussion of Schützenberger's jeu de taquin and related matters is included.

Chapter 4 recasts much of the previous work in the language of symmetric functions. Schur functions are introduced, first combinatorially as the generating functions for semistandard tableaux and then in terms of symmetric group characters. The chapter concludes with the famous Littlewood-Richardson and Murnaghan-Nakayama rules.

My debt to several other books will be evident. Much of Chapter 1 is based on Ledermann's exquisite text on group characters [Led 77]. Chapter 2 borrows heavily from the monograph of James [Jam 78], whereas Chapter 4 is inspired by Macdonald's already classic book [Mac 79]. Finally, the third chapter is a synthesis of material from the research literature.

There are numerous applications of representations of groups, and in particular of the symmetric group, to other areas. For example, they arise in physics [Boe 70], probability and statistics [Dia 88], topological graph theory [Whi 84], and the theory of partially ordered sets [Stn 82]. However, to keep the length of this text reasonable, I have discussed only the connections with combinatorial algorithms.

This book grew out of a course that I taught while visiting the Université du Québec à Montréal during the fall of 1986. I would like to thank *l'équipe de combinatoire* for arranging my stay. I also presented this material in a class here at Michigan State University in the winter and spring of 1990. I thank my students in both courses for many helpful suggestions (and those at l'UQAM for tolerating my bad French). Francesco Brenti, Kathy Dempsey, Yoav Dvir, Kathy Jankoviak, and Scott Mathison have all pointed out ways in which the presentation could be improved. I also wish to express my appreciation of John Kimmel, Marlene Thom, and Linda Loba at Wadsworth and Brooks/Cole for their help during the preparation of the manuscript. Because I typeset this document myself, all errors can be blamed on my computer.

Bruce E. Sagan

List of Symbols

Symbol	Meaning	page
e_t	polytabloid associated with Young tableau t	61
ev Q	the evacuation tableau of Q	128
ϵ	identity element of a group	1
f_S	weight generating function of weighted set S	141
f^λ	number of standard λ-tableau	73
ϕ^λ	character of permutation module M^λ	56
G	group	1
$g_{A,B}$	Garnir element of a pair of sets A, B	70
GL_d	$d \times d$ complex general linear group	4
GP	generalized permutations	165
GP$'$	generalized permutations, no repeated columns	168
h_n	n^{th} complete homogeneous symmetric function	148
h_λ	λ^{th} complete homogeneous symmetric function	150
Hom(V, W)	all module homomorphisms from V to W	22
$h_{i,j}$	hooklength of cell (i, j)	91
$H_{i,j}$	hook of cell (i, j)	91
h_v	hooklength of cell v	91
H_v	hook of cell v	91
I	identity matrix	23
I_d	$d \times d$ identity matrix	24
$i_k(\pi)$	length of π's longest k-increasing subsequence	109
im θ	image of θ	21
$j(P)$	jeu de taquin tableau of P	123
$j_c(P)$	forward slide on tableau P into cell c	120
$j^c(P)$	backward slide on tableau P into cell c	120
$j_Q(P)$	sequence of forward slides on P into Q	124
$j^P(Q)$	sequence of backward slides on Q into P	125
ker θ	kernel of θ	21
K_g	conjugacy class of a group element g	3
k_λ	size of K_λ	3
K_λ	conjugacy class in $\mathcal{S}_n$ corresponding to λ	3
$K_{\lambda\mu}$	Kostka number	85
κ_t	signed column sum for Young tableau t	61
$l(\lambda)$	length (number of parts) of partition λ	148
$ll(\xi)$	leg length of rim hook ξ	176

Symbol	Meaning	page
λ	partition	2
	composition	67
λ'	conjugate or transpose of partition λ	110
$(\lambda_1, \lambda_2, \ldots, \lambda_l)$	partition given as a sequence	2
	composition given as a sequence	15
$\lambda \vdash n$	λ is a partition of n	53
λ/μ	skew shape	119
$\lambda \backslash \xi$	partition λ with rim hook ξ removed	177
Λ	ring of symmetric functions	147
Λ_l	symmetric functions in l variables	159
Λ^n	symmetric functions homogeneous of degree n	148
Mat_d	full $d \times d$ matrix algebra	4
m_λ	λ^{th} monomial symmetric function	147
M^λ	permutation module corresponding to λ	56
P	partial tableau	93
$p(n)$	number of partitions of n	139
p_n	n^{th} power sum symmetric function	148
p_λ	λ^{th} power sum symmetric function	150
$P(\pi)$	P-tableau of permutation π	99
π	permutation	1
π_P	row word of the Young tableau P	107
π^r	reversal of permutation π	103
$\hat{\pi}$	top row of generalized permutation π	165
$\check{\pi}$	bottom row of generalized permutation π	165
$\pi \overset{\mathrm{R-S}}{\longleftrightarrow} (P, Q)$	Robinson-Schensted map	98
$\pi \overset{\mathrm{R-S-K}}{\longleftrightarrow} (P, Q)$	Robinson-Schensted-Knuth map	166
$\pi \overset{\mathrm{R-S-K}'}{\longleftrightarrow} (P, Q)$	dual Robinson-Schensted-Knuth map	168
$Q(\pi)$	Q-tableau of permutation π	99
$R(G)$	vector space of class functions on G	32
R^n	class functions of $\mathcal{S}_n$	163
R_t	row-stabilizer group of Young tableau t	60
r_x	row-insertion operator for the integer x	99
$\mathcal{S}_A$	group of permutations of A	20
s_λ	Schur function associated with λ	151
$s_{\lambda/\mu}$	skew Schur function associated with λ/μ	171
$\mathcal{S}_\lambda$	Young subgroup associated with partition λ	54
S^λ	Specht module associated with partition λ	62

Symbol	Meaning	page
$\mathrm{sgn}(\pi)$	sign of permutation π	4
$\mathrm{sh}\,t$	shape of Young tableau t	55
$\mathcal{S}_n$	symmetric group on $\{1, 2, \ldots, n\}$	1
t	Young tableau	55
T	generalized Young tableau	78
$\{t\}$	row tabloid of t	55
$[t]$	column tabloid of t	72
$\mathcal{T}_{\lambda\mu}$	generalized tableaux, shape λ, content μ	78
$\mathcal{T}_{\lambda\mu}^0$	semistandard tableaux, shape λ, content μ	81
θ	homomorphism of modules	18
θ_T	homomorphism corresponding to tableau T	80
$\bar{\theta}_T$	restriction of θ_T to a Specht module	81
$v_Q(P)$	vacating tableau for $j_Q(P)$	124
$v^P(Q)$	vacating tableau for $j^P(Q)$	125
$W^\perp$	orthogonal complement of W	15
wt	weight function	141
$\mathbf{x}$	the set of variables $\{x_1, x_2, x_3, \ldots\}$	147
$X(g)$	matrix of g in the representation X	4
$\mathbf{x}^T$	monomial weight of a tableau T	151
$\mathbf{x}^\mu$	monomial weight of a composition μ	151
$X{\downarrow}_H^G$	restriction of representation X from G to H	45
$X{\uparrow}_H^G$	induction of representation X from H to G	45
ξ	rim or skew hook	176
Z_A	center of algebra A	27
Z_g	centralizer of group element g	3
Z_λ	centralizer in $\mathcal{S}_n$ corresponding to λ	3
z_λ	size of Z_λ	3
1_G	trivial representation of G	4
$(1^{m_1}, 2^{m_2}, \ldots)$	partition given by multiplicities	2
$\cong$	equivalence of modules	19
	slide equivalence of tableaux	121
$\overset{*}{\cong}$	dual equivalence of tableaux	124
$\overset{K}{\cong}$	Knuth equivalence	106
$\overset{K^*}{\cong}$	dual Knuth equivalence	118
$\overset{P}{\cong}$	P-equivalence	105
$\overset{Q}{\cong}$	Q-equivalence	118

The Symmetric Group

Representations, Combinatorial Algorithms, and Symmetric Functions

Chapter 1

Group Representations

We begin our study of the symmetric group by considering its representations. First, however, we must present some general results about group representations that will be useful in our special case. Representation theory can be couched in terms of matrices or in the language of modules. We consider both approaches and then turn to the associated theory of characters. All our work will use the complex numbers as the ground field in order to make life as easy as possible.

We are presenting the material in this chapter so that this book will be relatively self-contained, although it all can be found in other standard texts. In particular, our exposition is modeled on the one in Ledermann [Led 77].

1.1 Fundamental Concepts

In this section we introduce some basic terminology and notation. We pay particular attention to the symmetric group.

Let G be a group written multiplicatively with identity ϵ. Throughout this work, G is finite unless stated otherwise. We assume the reader is familiar with the elementary properties of groups (cosets, Lagrange's theorem, etc.) that can be found in any standard text such as Herstein [Her 64].

Our object of study is the *symmetric group*, $\mathcal{S}_n$, consisting of all bijections from $\{1, 2, \ldots, n\}$ to itself using composition as the multiplication. The elements $\pi \in \mathcal{S}_n$ are called *permutations*. We multiply permutations from right to left. (In fact, we compose all functions in this manner.) Thus $\pi\sigma$ is the bijection obtained by first applying σ, followed by π.

If π is a permutation, then there are three different notations we can use for this element. *Two-line notation* is the array

$$
\begin{matrix}
1 & 2 & \cdots & n \\
\pi(1) & \pi(2) & \cdots & \pi(n)
\end{matrix}
$$

1

For example, if $\pi \in \mathcal{S}_5$ is given by

$$\pi(1) = 2, \qquad \pi(2) = 3, \qquad \pi(3) = 1, \qquad \pi(4) = 4, \qquad \pi(5) = 5$$

then its two-line form is

$$\pi = \begin{array}{ccccc} 1 & 2 & 3 & 4 & 5 \\ 2 & 3 & 1 & 4 & 5 \end{array}$$

Because the top line is fixed, we can drop it to get *one-line notation*.

Lastly, we can display π using *cycle notation*. Given $i \in \{1, 2, \ldots, n\}$, the elements of the sequence $i, \pi(i), \pi^2(i), \pi^3(i), \ldots$ cannot all be distinct. Taking the first power p such that $\pi^p(i) = i$, we have the cycle

$$(i, \pi(i), \pi^2(i), \ldots, \pi^{p-1}(i))$$

Equivalently, the cycle $(i, j, k, \ldots, l)$ means π sends i to j, j to k, $\ldots$, and l back to i. Now pick an element not in the cycle containing i and iterate this process until all members of $\{1, 2, \ldots, n\}$ have been used. Our example from the last paragraph becomes

$$\pi = (1, 2, 3)(4)(5)$$

in cycle notation. Note that cyclically permuting the elements within a cycle or reordering the cycles themselves does not change the permutation. Thus

$$(1, 2, 3)(4)(5) = (2, 3, 1)(4)(5) = (4)(2, 3, 1)(5) = (4)(5)(3, 1, 2)$$

A *k-cycle*, or *cycle of length k*, is a cycle containing k elements. The preceding permutation consists of a 3-cycle and two 1-cycles. The *cycle type*, or simply the *type*, of π is an expression of the form

$$(1^{m_1}, 2^{m_2}, \ldots, n^{m_n})$$

where m_k is the number of cycles of length k in π. The example permutation has cycle type

$$(1^2, 2^0, 3^1, 4^0, 5^0)$$

A 1-cycle of π is called a *fixed-point*. The numbers 4 and 5 are fixed-points in our example. Fixed-points are usually dropped from the cycle notation if no confusion will result. An *involution* is a permutation such that $\pi^2 = \epsilon$. It is easy to see that π is an involution if and only if all of π's cycles have length 1 or 2.

Another way to give the cycle type is as a partition. A *partition of n* is a sequence

$$\lambda = (\lambda_1, \lambda_2, \ldots, \lambda_l)$$

where the λ_i are weakly decreasing and $\sum_{i=1}^{l} \lambda_i = n$. Thus k is repeated m_k times in the partition version of the cycle type of π. Our example corresponds to the partition

$$\lambda = (3, 1, 1)$$

In any group G, elements g and h are *conjugates* if

$$g = khk^{-1}$$

for some $k \in G$. The set of all elements conjugate to a given g is called the *conjugacy class of g* and is denoted K_g. Conjugacy is an equivalence relation, so the distinct conjugacy classes partition G. (This is a *set* partition, as opposed to the *integer* partitions discussed in the previous paragraph.) Returning to $\mathcal{S}_n$, it is not hard to see that if

$$\pi = (i_1, i_2, \ldots, i_l) \ldots (i_m, i_{m+1}, \ldots, i_n)$$

in cycle notation, then for any $\sigma \in \mathcal{S}_n$

$$\sigma \pi \sigma^{-1} = (\sigma(i_1), \sigma(i_2), \ldots, \sigma(i_l)) \ldots (\sigma(i_m), \sigma(i_{m+1}), \ldots, \sigma(i_n))$$

It follows that two permutations are in the same conjugacy class if and only if they have the same cycle type. Thus there is a natural one-to-one correspondence between partitions of n and conjugacy classes of $\mathcal{S}_n$.

We can compute the size of a conjugacy class in the following manner. Let G be any group and consider the *centralizer* of $g \in G$ defined by

$$Z_g = \{h \in G \mid hgh^{-1} = g\}$$

—i.e., the set of all elements that commute with g. Now there is a bijection between the cosets of Z_g and the elements of K_g, so that

$$|K_g| = \frac{|G|}{|Z_g|} \tag{1.1}$$

where $|\cdot|$ denotes cardinality. Now let $G = \mathcal{S}_n$ and use K_λ, Z_λ for K_g, Z_g when g has type λ.

Proposition 1.1.1 *If $\lambda = (1^{m_1}, 2^{m_2}, \ldots, n^{m_n})$, then*

$$z_\lambda \overset{\text{def}}{=} |Z_\lambda| = 1^{m_1} m_1! 2^{m_2} m_2! \cdots n^{m_n} m_n!$$

Proof. Any $h \in Z_\lambda$ can either permute the cycles of length i among themselves or perform a cyclic rotation on each of the individual cycles (or both). Since there are $m_i!$ ways to do the former operation and i^{m_i} ways to do the latter, we are done. ■

Thus equation (1.1) specializes in the symmetric group to

$$k_\lambda = \frac{n!}{z_\lambda} = \frac{n!}{1^{m_1} m_1! 2^{m_2} m_2! \cdots n^{m_n} m_n!} \tag{1.2}$$

where $k_\lambda = |K_\lambda|$.

Of particular interest is the conjugacy class of *transpositions*, which are those permutations of the form $\tau = (i, j)$. The transpositions generate $\mathcal{S}_n$ as a group; in fact the symmetric group is generated by the *adjacent transpositions* $(1, 2)$, $(2, 3)$, ..., $(n-1, n)$. If $\pi = \tau_1 \tau_2 \cdots \tau_k$, where the τ_i are transpositions, then we define the *sign of* π to be

$$\text{sgn}(\pi) = (-1)^k$$

It can be proved that sgn is well defined, i.e., independent of the particular decomposition of π into transpositions. Once this is established, it follows easily that

$$\text{sgn}(\pi\sigma) = \text{sgn}(\pi)\,\text{sgn}(\sigma) \tag{1.3}$$

As we will see, this is an example of a representation.

1.2   Matrix Representations

A matrix representation can be thought of as a way to model an abstract group with a concrete group of matrices. After giving the precise definition, we look at some examples.

Let $\mathbf{C}$ denote the complex numbers. Let Mat_d stand for the set of all $d \times d$ matrices with entries in $\mathbf{C}$. This is called a *full complex matrix algebra of degree d*. Recall that an algebra is a vector space with an associative multiplication of vectors (thus also imposing a ring structure on the space). The *complex general linear group of degree d*, denoted GL_d, is the group of all $X = (x_{i,j})_{d \times d} \in \text{Mat}_d$ that are invertible with respect to multiplication.

Definition 1.2.1 A *matrix representation of a group G* is a group homomorphism

$$X : G \to GL_d$$

Equivalently, to each $g \in G$ is assigned $X(g) \in \text{Mat}_d$ such that

1. $X(\epsilon) = I$ the identity matrix, and

2. $X(gh) = X(g)X(h)$ for all $g, h \in G$.

The parameter d is called the *degree*, or *dimension*, of the representation and is denoted $\deg X$. ∎

Note that conditions 1 and 2 imply that $X(g^{-1}) = X(g)^{-1}$, so these matrices are in GL_d as required.

Obviously the simplest representations are those of degree 1. Our first two examples are of this type.

Example 1.2.2 All groups have the *trivial representation*, which is the one sending every $g \in G$ to the matrix (1). This is clearly a representation because $X(\epsilon) = (1)$ and

$$X(g)X(h) = (1)(1) = (1) = X(gh)$$

for all $g, h \in G$. We often use 1_G or just the number 1 itself to stand for the trivial representation of G. ∎

Example 1.2.3 Let us find all one-dimensional representations of the cyclic group of order n, C_n. Let g be a generator of C_n, i.e.,

$$C_n = \{g, g^2, g^3, \ldots, g^n = \epsilon\}$$

If $X(g) = (c), c \in \mathbf{C}$, then the matrix for every element of C_n is determined since $X(g^k) = (c^k)$ by property 2 in the preceding definition. But by property 1,

$$(c^n) = X(g^n) = X(\epsilon) = (1)$$

so c must be an nth root of unity. Clearly each such root gives a representation, so there are exactly n representations of C_n having degree 1.

In particular, let $n = 4$ and $C_4 = \{\epsilon, g, g^2, g^3\}$. The four fourth roots of unity are $1, i, -1, -i$. If we let the four corresponding representations be denoted $X^{(1)}, X^{(2)}, X^{(3)}, X^{(4)}$, then we can construct a table:

	ϵ	g	g^2	g^3
$X^{(1)}$	1	1	1	1
$X^{(2)}$	1	i	-1	$-i$
$X^{(3)}$	1	-1	1	-1
$X^{(4)}$	1	$-i$	-1	i

where the entry in row i and column j is $X^{(i)}(g^j)$ (matrix brackets omitted). This array is an example of a character table, a concept we will develop in Section 1.8. (For representations of dimension 1, the representation and its character coincide.) Note that the trivial representation forms the first row of the table.

There are other representations of C_4 of larger degree. For example, we can let

$$X(g) = \begin{pmatrix} 1 & 0 \\ 0 & i \end{pmatrix}$$

However, this representation is really just a combination of $X^{(1)}$ and $X^{(2)}$. In the language of Section 1.5, X is completely reducible with irreducible components $X^{(1)}$ and $X^{(2)}$. We will see that every representation of C_n can be constructed in this way using the n degree 1 representations as building blocks. ∎

Example 1.2.4 We have already met a nontrivial degree 1 representation of $\mathcal{S}_n$. In fact equation (1.3) merely says that the map $X(\pi) = (\text{sgn}(\pi))$ is a representation called the *sign representation*.

Also of importance is the *defining representation* of $\mathcal{S}_n$, which is of degree n. If $\pi \in \mathcal{S}_n$, then we let $X(\pi) = (x_{i,j})_{n \times n}$, where

$$x_{i,j} = \begin{cases} 1 & \text{if } \pi(j) = i \\ 0 & \text{otherwise} \end{cases}$$

The matrix $X(\pi)$ is called a *permutation matrix*, since it contains only zeros and ones, with a unique one in every row and column. The reader should verify that this is a representation.

In particular, consider $\mathcal{S}_3$ with its permutations written in cycle notation. Then the matrices of the defining representation are

$$X(\epsilon) = \begin{pmatrix} 1 & 0 & 0 \\ 0 & 1 & 0 \\ 0 & 0 & 1 \end{pmatrix}, \quad X(\,(1,2)\,) = \begin{pmatrix} 0 & 1 & 0 \\ 1 & 0 & 0 \\ 0 & 0 & 1 \end{pmatrix}$$

$$X(\,(1,3)\,) = \begin{pmatrix} 0 & 0 & 1 \\ 0 & 1 & 0 \\ 1 & 0 & 0 \end{pmatrix}, \quad X(\,(2,3)\,) = \begin{pmatrix} 1 & 0 & 0 \\ 0 & 0 & 1 \\ 0 & 1 & 0 \end{pmatrix}$$

$$X(\,(1,2,3)\,) = \begin{pmatrix} 0 & 0 & 1 \\ 1 & 0 & 0 \\ 0 & 1 & 0 \end{pmatrix}, \quad X(\,(1,3,2)\,) = \begin{pmatrix} 0 & 1 & 0 \\ 0 & 0 & 1 \\ 1 & 0 & 0 \end{pmatrix} \blacksquare$$

1.3 G-modules and the Group Algebra

Because matrices correspond to linear transformations, we can think of representations in these terms. This is the idea of a G-module.

Let V be a vector space. Unless stated otherwise, all vector spaces will be over the complex numbers and of finite dimension. Let $GL(V)$ stand for the set of all invertible linear transformations of V to itself, called the *general linear group of V*. If $\dim V = d$, then $GL(V)$ and GL_d are isomorphic as groups.

Definition 1.3.1 Let V be a vector space and G be a group. Then V is a *G-module* if there is a group homomorphism

$$\rho : G \to GL(V)$$

Equivalently, V is a G-module if there is a multiplication, $g\mathbf{v}$, of elements of V by elements of G such that

1. $g\mathbf{v} \in V$,

2. $g(c\mathbf{v} + d\mathbf{w}) = c(g\mathbf{v}) + d(g\mathbf{w})$,

3. $(gh)\mathbf{v} = g(h\mathbf{v})$, and

4. $\epsilon\mathbf{v} = \mathbf{v}$

for all $g, h \in G$; $\mathbf{v}, \mathbf{w} \in V$ and scalars $c, d \in \mathbf{C}$. $\blacksquare$

In the future, G-module will often be shortened to module when no confusion can result about the group involved. Other words involving G- as a prefix will be treated similarly. Alternatively, we can say that the vector space V *carries* a representation of G.

Let us verify that the two parts of the definition are equivalent. In fact, we are just using $g\mathbf{v}$ as a shorthand for the application of the transformation $\rho(g)$ to the vector $\mathbf{v}$. Item 1 says that the transformation takes V to itself; 2 shows the map is linear; 3 is property 2 of the matrix definition; and 4 in combination with 3 says that g and g^{-1} are inverse maps, so all transformations are invertible. Although it is more abstract than our original definition of representation, the G-module concept lends itself to cleaner proofs.

In fact, we can go back and forth between these two notions of representation quite easily. Given a matrix representation X of degree d, let V be the vector space $\mathbf{C}^d$ of all column vectors of length d. Then we can multiply $\mathbf{v} \in V$ by $g \in G$ using the definition

$$g\mathbf{v} \overset{\text{def}}{=} X(g)\mathbf{v}$$

where the operation on the right is matrix multiplication. Conversely, if V is a G-module, then take any basis $\mathcal{B}$ for V. Thus $X(g)$ will just be the matrix of the linear transformation $g \in G$ in the basis $\mathcal{B}$ computed in the usual way. We use this correspondence extensively in the rest of this book.

Note that if S is any set with a multiplication by elements of G satisfying 1, 3, and 4, then we say G *acts on* S. Group actions are important in their own right. In fact, it is always possible to take a set on which G acts and turn it into a G-module as follows. Let $\mathbf{C}[S]$ denote the vector space generated by S over $\mathbf{C}$; i.e., $\mathbf{C}[S]$ consists of all the formal linear combinations

$$c_1 \mathbf{s}_1 + c_2 \mathbf{s}_2 + \cdots + c_n \mathbf{s}_n$$

where $c_i \in \mathbf{C}$ for all i and $S = \{s_1, s_2, \ldots, s_n\}$. (We put the elements of S in boldface print when they are being considered as vectors.) Vector addition and scalar multiplication in $\mathbf{C}[S]$ are defined by

$$
\begin{aligned}
&(c_1 \mathbf{s}_1 + c_2 \mathbf{s}_2 + \cdots + c_n \mathbf{s}_n) \ + \ (d_1 \mathbf{s}_1 + d_2 \mathbf{s}_2 + \cdots + d_n \mathbf{s}_n) \\
&= \ (c_1 + d_1)\mathbf{s}_1 + (c_2 + d_2)\mathbf{s}_2 + \cdots + (c_n + d_n)\mathbf{s}_n
\end{aligned}
$$

and

$$c\,(c_1 \mathbf{s}_1 + c_2 \mathbf{s}_2 + \cdots + c_n \mathbf{s}_n) = (cc_1)\mathbf{s}_1 + (cc_2)\mathbf{s}_2 + \cdots + (cc_n)\mathbf{s}_n$$

respectively. Now the action of G on S can be extended to an action on $\mathbf{C}[S]$ by linearity:

$$g\,(c_1 \mathbf{s}_1 + c_2 \mathbf{s}_2 + \cdots + c_n \mathbf{s}_n) = c_1(g\mathbf{s}_1) + c_2(g\mathbf{s}_2) + \cdots + c_n(g\mathbf{s}_n)$$

for all $g \in G$. This makes $\mathbf{C}[S]$ into a G-module of dimension $|S|$.

Definition 1.3.2 If a group G acts on a set S, then the associated module $\mathbf{C}[\mathbf{S}]$ is called the *permutation representation* associated with S. Also, the elements of S form a basis for $\mathbf{C}[\mathbf{S}]$ called the *standard basis*. ∎

All the following examples of G-modules are of this form.

Example 1.3.3 Consider the symmetric group $\mathcal{S}_n$ with its usual action on $S = \{1, 2, \ldots, n\}$. Now

$$\mathbf{C}[\mathbf{S}] = \{c_1\mathbf{1} + c_2\mathbf{2} + \cdots + c_n\mathbf{n} \,|\, c_i \in \mathbf{C} \text{ for all } i\}$$

with the action

$$\pi(c_1\mathbf{1} + c_2\mathbf{2} + \cdots + c_n\mathbf{n}) = c_1\pi(\mathbf{1}) + c_2\pi(\mathbf{2}) + \cdots + c_n\pi(\mathbf{n})$$

for all $\pi \in \mathcal{S}_n$.

To make things more concrete, we can select a basis and determine the matrices $X(\pi)$ for $\pi \in \mathcal{S}_n$ in that basis. Let us consider $\mathcal{S}_3$ and use the standard basis $\{\mathbf{1}, \mathbf{2}, \mathbf{3}\}$. To find the matrix for $\pi = (1, 2)$, we compute

$$(1, 2)\mathbf{1} = \mathbf{2}; \qquad (1, 2)\mathbf{2} = \mathbf{1}; \qquad (1, 2)\mathbf{3} = \mathbf{3}$$

and so

$$X(\,(1, 2)\,) = \begin{pmatrix} 0 & 1 & 0 \\ 1 & 0 & 0 \\ 0 & 0 & 1 \end{pmatrix}$$

If the reader determines the rest of the matrices for $\mathcal{S}_3$, it will be noted that they are exactly the same as those of the defining representation, Example 1.2.4. It is not hard to show that the same is true for any n; i.e., this is merely the module approach to the defining representation of $\mathcal{S}_n$. ∎

Example 1.3.4 We now describe one of the most important representations for any group, the *(left) regular representation*. Let G be an arbitrary group. Then G acts on itself by left multiplication: if $g \in G$ and $h \in S = G$, then the action of g on h, gh, is defined as the usual product in the group. Properties 1, 3, and 4 now follow, respectively, from the closure, associativity, and identity axioms for the group.

Thus if $G = \{g_1, g_2, \ldots, g_n\}$, then we have the corresponding G-module

$$\mathbf{C}[\mathbf{G}] = \{c_1\mathbf{g_1} + c_2\mathbf{g_2} + \cdots + c_n\mathbf{g_n} \,|\, c_i \in \mathbf{C} \text{ for all } i\}$$

which is called the *group algebra of* G. The action is, of course,

$$g\,(c_1\mathbf{g_1} + c_2\mathbf{g_2} + \cdots + c_n\mathbf{g_n}) = c_1(\mathbf{gg_1}) + c_2(\mathbf{gg_2}) + \cdots + c_n(\mathbf{gg_n})$$

for all $g \in G$. The group algebra will furnish us with much important combinatorial information about group representations.

Let's see what the regular representation of the cyclic group C_4 looks like. First of all

$$\mathbf{C}[\mathbf{C_4}] = \{c_1\epsilon + c_2\mathbf{g} + c_3\mathbf{g}^2 + c_4\mathbf{g}^3 \mid c_i \in \mathbf{C} \text{ for all } i\}$$

We can easily find the matrix of g^2 in the standard basis:

$$g^2\epsilon = \mathbf{g}^2, \qquad g^2\mathbf{g} = \mathbf{g}^3, \qquad g^2\mathbf{g}^2 = \epsilon, \qquad g^2\mathbf{g}^3 = \mathbf{g}$$

Thus

$$X(g^2) = \begin{pmatrix} 0 & 0 & 1 & 0 \\ 0 & 0 & 0 & 1 \\ 1 & 0 & 0 & 0 \\ 0 & 1 & 0 & 0 \end{pmatrix}$$

Computing the rest of the matrices would be a good exercise. Note that they are all permutation matrices and all distinct. In general, the regular representation of G gives an embedding of G into the symmetric group on $|G|$ elements. The reader has probably seen this presented in a group theory course, in a slightly different guise, as Cayley's theorem [Her 64, pages 60–61].

Note that if G acts on any V, then so does $\mathbf{C}[\mathbf{G}]$. Specifically, if $c_1\mathbf{g}_1 + c_2\mathbf{g}_2 + \cdots + c_n\mathbf{g}_n \in \mathbf{C}[\mathbf{G}]$ and $\mathbf{v} \in V$, then we can define the action

$$(c_1\mathbf{g}_1 + c_2\mathbf{g}_2 + \cdots + c_n\mathbf{g}_n)\mathbf{v} = c_1(g_1\mathbf{v}) + c_2(g_2\mathbf{v}) + \cdots + c_n(g_n\mathbf{v})$$

In fact, we can extend the concept of representation to algebras: a representation of an algebra A is an algebra homomorphism from A into $GL(V)$. In this way, every representation of a group G gives rise to a representation of its group algebra $\mathbf{C}[\mathbf{G}]$. For a further discussion of representations of algebras, see the text of Curtis and Reiner [C-R 66]. ∎

Example 1.3.5 Let group G have subgroup H, written $H \leq G$. A generalization of the regular representation is the *(left) coset representation* of G with respect to H. Let $g_1, g_2, \ldots, g_k$ be a transversal for H; i.e., $\mathcal{H} = \{g_1H, g_2H, \ldots, g_kH\}$ is a complete set of disjoint left cosets for H in G. Then G acts on $\mathcal{H}$ by letting

$$g(g_iH) = (gg_i)H$$

for all $g \in G$. The corresponding module

$$\mathbf{C}[\mathcal{H}] = \{c_1\mathbf{g}_1\mathbf{H} + c_2\mathbf{g}_2\mathbf{H} + \cdots + c_k\mathbf{g}_k\mathbf{H} \mid c_i \in \mathbf{C} \text{ for all } i\}$$

inherits the action

$$g(c_1\mathbf{g}_1\mathbf{H} + \cdots + c_k\mathbf{g}_k\mathbf{H}) = c_1(\mathbf{gg}_1\mathbf{H}) + \cdots + c_k(\mathbf{gg}_k\mathbf{H})$$

Note that if $H = G$, then this reduces to the trivial representation. At the other extreme, when $H = \{\epsilon\}$, then $\mathcal{H} = G$ and we obtain the regular representation again. In general, representation by cosets is an example of an induced representation, an idea studied further in Section 1.12.

Let us consider $G = \mathcal{S}_3$ and $H = \{\epsilon, (2,3)\}$. We can take

$$\mathcal{H} = \{H,\ (1,2)H,\ (1,3)H\}$$

and

$$\mathbf{C}[\mathcal{H}] = \{c_1\mathbf{H} + c_2(\mathbf{1,2})\mathbf{H} + c_3(\mathbf{1,3})\mathbf{H} \mid c_i \in \mathbf{C} \text{ for all } i\}$$

Computing the matrix of $(1,2)$ in the standard basis, we obtain

$$(\mathbf{1,2})\mathbf{H} = (\mathbf{1,2})\mathbf{H},\ (\mathbf{1,2})(\mathbf{1,2})\mathbf{H} = \mathbf{H},\ (\mathbf{1,2})(\mathbf{1,3})\mathbf{H} = (\mathbf{1,3,2})\mathbf{H} = (\mathbf{1,3})\mathbf{H}$$

so that

$$X(\ (1,2)\ ) = \begin{pmatrix} 0 & 1 & 0 \\ 1 & 0 & 0 \\ 0 & 0 & 1 \end{pmatrix}$$

After finding a few more matrices, the reader will become convinced that we have rediscovered the defining representation yet again. The reason for this is explained when we consider isomorphism of modules in Section 1.6. ∎

1.4   Reducibility

An idea pervading all of science is that large structures can be understood by breaking them up into their smallest pieces. The same thing is true of representation theory. Some representations are built out of smaller ones (such as the one at the end of Example 1.2.3), wheras others are indivisible (as are all degree one representations). This is the distinction between reducible and irreducible representations, which we study in this section. First, however, we must determine precisely what a piece or subobject means in this setting.

Definition 1.4.1 Let V be a G-module. A *submodule of V* is a subspace W that is closed under the action of G—i.e.,

$$\mathbf{w} \in W \Rightarrow g\mathbf{w} \in W \text{ for all } g \in G$$

We also say that W is a *G-invariant subspace*. Equivalently, W is a subset of V that is a G-module in its own right. We write $W \leq V$ if W is a submodule of V. ∎

As usual, we illustrate the definition with some examples.

Example 1.4.2 Any G-module, V, has the submodules $W = V$ as well as $W = \{\mathbf{0}\}$, where $\mathbf{0}$ is the zero vector. These two submodules are called *trivial*. All other submodules are called *nontrivial*. ∎

Example 1.4.3 For a nontrivial example of a submodule, consider $G = \mathcal{S}_n$, $n \geq 2$, and $V = \mathbf{C}[\mathbf{1}, \mathbf{2}, \ldots, \mathbf{n}]$ (the defining representation). Now take

$$W = \mathbf{C}[\mathbf{1} + \mathbf{2} + \cdots + \mathbf{n}] = \{c(\mathbf{1} + \mathbf{2} + \cdots + \mathbf{n}) \mid c \in \mathbf{C}\}$$

—i.e., W is the one-dimensional subspace spanned by the vector $\mathbf{1}+\mathbf{2}+\cdots+\mathbf{n}$. To check that W is closed under the action of $\mathcal{S}_n$, it suffices to show that

$$\pi \mathbf{w} \in W \text{ for all } \mathbf{w} \text{ in some basis for } W \text{ and all } \pi \in \mathcal{S}_n$$

(Why?) Thus we need to verify only that

$$\pi(\mathbf{1} + \mathbf{2} + \cdots + \mathbf{n}) \in W$$

for each $\pi \in \mathcal{S}_n$. But

$$\begin{aligned}
\pi(\mathbf{1} + \mathbf{2} + \cdots + \mathbf{n}) &= \pi(\mathbf{1}) + \pi(\mathbf{2}) + \cdots + \pi(\mathbf{n}) \\
&= \mathbf{1} + \mathbf{2} + \cdots + \mathbf{n} \ \in W
\end{aligned}$$

because applying π to $\{1, 2, \ldots, n\}$ just gives back the same set of numbers in a different order. Thus W is a submodule of V that is nontrivial since $\dim W = 1$ and $\dim V = n \geq 2$.

Since W is a module for G sitting inside V, we can ask what representation we get if we restrict the action of G to W. But we have just shown that every $\pi \in \mathcal{S}_n$ sends the basis vector $\mathbf{1} + \mathbf{2} + \cdots + \mathbf{n}$ to itself. Thus $X(\pi) = (1)$ is the corresponding matrix, and we have found a copy of the trivial representation in $\mathbf{C}[\mathbf{1}, \mathbf{2}, \ldots, \mathbf{n}]$. . In general, for a vector space W of any dimension, if G fixes every element of W, we say that G *acts trivially on* W. ∎

Example 1.4.4 Next, let's look again at the regular representation. Suppose $G = \{g_1, g_2, \ldots, g_n\}$ with group algebra $V = \mathbf{C}[\mathbf{G}]$. Using the same idea as in the previous example, let

$$W = \mathbf{C}[\mathbf{g}_1 + \mathbf{g}_2 + \cdots + \mathbf{g}_n]$$

the one-dimensional subspace spanned by the vector that is the sum of all the elements of G. To verify that W is a submodule, take any $g \in G$ and compute:

$$\begin{aligned}
g(\mathbf{g}_1 + \mathbf{g}_2 + \cdots + \mathbf{g}_n) &= g\mathbf{g}_1 + g\mathbf{g}_2 + \cdots + g\mathbf{g}_n \\
&= \mathbf{g}_1 + \mathbf{g}_2 + \cdots + \mathbf{g}_n \ \in W
\end{aligned}$$

because multiplying by g merely permutes the elements of G, leaving the sum unchanged. As before, G acts trivially on W.

The reader should verify that if $G = \mathcal{S}_n$, then the sign representation can also be recovered by using the submodule

$$W = \mathbf{C}[\textstyle\sum_{\pi \in \mathcal{S}_n} \operatorname{sgn}(\pi) \ \pi \ ] \ \blacksquare$$

We now introduce the irreducible representations which will be the building blocks of all the others.

Definition 1.4.5 A nonzero G-module V is *reducible* if it contains a nontrivial submodule W. Otherwise, V is said to be *irreducible*. Equivalently, V is reducible if it has a basis $\mathcal{B}$ in which every $g \in G$ is assigned a block matrix of the form

$$X(g) = \left(\begin{array}{c|c} A(g) & B(g) \\ \hline 0 & C(g) \end{array} \right) \tag{1.4}$$

where the $A(g)$ are square matrices, all of the same size, and 0 is a nonempty matrix of zeros. ∎

To see the equivalence, suppose V of dimension d has a submodule W of dimension f, $0 < f < d$. Then let

$$\mathcal{B} = \{\mathbf{w}_1, \mathbf{w}_2, \ldots, \mathbf{w}_f, \mathbf{v}_{f+1}, \mathbf{v}_{f+2}, \ldots, \mathbf{v}_d\}$$

where the first f vectors are a basis for W. Now we can compute the matrix of any $g \in G$ with respect to the basis $\mathcal{B}$. Since W is a submodule, $g\mathbf{w}_i \in W$ for all i, $1 \leq i \leq f$. Thus the last $d - f$ coordinates of $g\mathbf{w}_i$ will all be zero. That accounts for the zero matrix in the lowerleft corner of $X(g)$. Note that we have also shown that the $A(g)$, $g \in G$, are the matrices of the restriction of G to W. Hence they must all be square and of the same size.

Conversely, suppose each $X(g)$ has the given form with every $A(g)$ being $f \times f$. Let $V = \mathbf{C}^d$ and consider

$$W = \mathbf{C}[\mathbf{e}_1, \mathbf{e}_2, \ldots, \mathbf{e}_f]$$

where $\mathbf{e}_i$ is the column vector with a 1 in the ith row and zeros elsewhere (the *standard basis* for $\mathbf{C}^d$). Then the placement of the zeros in $X(g)$ assures us that $X(g)\mathbf{e}_i \in W$ for $1 \leq i \leq f$ and all $g \in G$. Thus W is a G-module, and it is nontrivial because the matrix of zeros is nonempty.

Clearly any degree 1 representation is irreducible. It seems hard to determine when a representation of greater degree will be irreducible. Certainly, checking all possible subspaces to find out which ones are submodules is out of the question. This unsatisfactory state of affairs will be remedied after we discuss inner products of group characters in Section 1.9.

From the preceding examples, both the defining representation for $\mathcal{S}_n$ and the group algebra for an arbitrary G are reducible if $n \geq 2$ and $|G| \geq 2$, respectively. After all, we produced nontrivial submodules. Let us now illustrate the alternative approach via matrices using the defining representation of $\mathcal{S}_3$. We must extend the basis $\{\mathbf{1}+\mathbf{2}+\mathbf{3}\}$ for W to a basis for $V = \mathbf{C}[\mathbf{1}, \mathbf{2}, \mathbf{3}]$. Let's pick

$$\mathcal{B} = \{\mathbf{1} + \mathbf{2} + \mathbf{3}, \ \mathbf{2}, \ \mathbf{3}\}$$

Of course $X(e)$ remains the 3×3 identity matrix. To compute $X(\ (1,2)\ )$, we look at $(1,2)$'s action on our basis:

$$(1,2)(\mathbf{1} + \mathbf{2} + \mathbf{3}) = \mathbf{1} + \mathbf{2} + \mathbf{3}, \ (1,2)\mathbf{2} = \mathbf{1} = (\mathbf{1} + \mathbf{2} + \mathbf{3}) - \mathbf{2} - \mathbf{3}, \ (1,2)\mathbf{3} = \mathbf{3}$$

So

$$X(\,(1,2)\,) = \begin{pmatrix} 1 & 1 & 0 \\ 0 & -1 & 0 \\ 0 & -1 & 1 \end{pmatrix}$$

The reader can do the similar computations for the remaining four elements of $\mathcal{S}_3$ to verify that

$$X(\,(1,3)\,) = \begin{pmatrix} 1 & 0 & 1 \\ 0 & 1 & -1 \\ 0 & 0 & -1 \end{pmatrix}, \qquad X(\,(2,3)\,) = \begin{pmatrix} 1 & 0 & 0 \\ 0 & 0 & 1 \\ 0 & 1 & 0 \end{pmatrix}$$

$$X(\,(1,2,3)\,) = \begin{pmatrix} 1 & 0 & 1 \\ 0 & 0 & -1 \\ 0 & 1 & -1 \end{pmatrix}, \qquad X(\,(1,3,2)\,) = \begin{pmatrix} 1 & 1 & 0 \\ 0 & -1 & 1 \\ 0 & -1 & 0 \end{pmatrix}$$

Note that all these matrices have the form

$$X(\pi) = \left(\begin{array}{c|cc} 1 & * & * \\ \hline 0 & * & * \\ 0 & * & * \end{array} \right)$$

The one in the upperleft corner comes from the fact that $\mathcal{S}_3$ acts trivially on W.

1.5 Complete Reducibility and Maschke's Theorem

It would be even better if we could bring the matrices of a reducible G-module to the block diagonal form

$$X(g) = \left(\begin{array}{c|c} A(g) & 0 \\ \hline 0 & B(g) \end{array} \right)$$

for all $g \in G$. This is the notion of a direct sum.

Definition 1.5.1 Let V be a vector space with subspaces U and W. Then V is the *(internal) direct sum of U and W*, written $V = U \oplus W$, if every $\mathbf{v} \in V$ can be written uniquely as a sum

$$\mathbf{v} = \mathbf{u} + \mathbf{w} \qquad \mathbf{u} \in U,\ \mathbf{w} \in W$$

If V is a G-module and U, W are G-submodules, then we say that U *and* W *are complements of each other*.

If X is a matrix, then X is the *direct sum of matrices A and B*, written $X = A \oplus B$, if X has the block diagonal form

$$X = \left(\begin{array}{c|c} A & 0 \\ \hline 0 & B \end{array} \right) \quad \blacksquare$$

To see the relationship between the module and matrix definitions, let V be a G-module with $V = U \oplus W$, where $U, W \leq V$. Since this is a direct sum of vector spaces, we can choose a basis for V

$$\mathcal{B} = \{\mathbf{u}_1, \mathbf{u}_2, \ldots, \mathbf{u}_f, \mathbf{w}_{f+1}, \mathbf{w}_{f+2}, \ldots, \mathbf{w}_d\}$$

such that $\{\mathbf{u}_1, \mathbf{u}_2, \ldots, \mathbf{u}_f\}$ is a basis for U and $\{\mathbf{w}_{f+1}, \mathbf{w}_{f+2}, \ldots, \mathbf{w}_d\}$ is a basis for W. Since U and W are submodules, we have

$$g\mathbf{u}_i \in U \quad \text{and} \quad g\mathbf{w}_j \in W$$

for all $g \in G$, $\mathbf{u}_i \in U$, $\mathbf{w}_j \in W$. Thus the matrix of any $g \in G$ in the basis $\mathcal{B}$ is

$$X(g) = \left(\begin{array}{c|c} A(g) & 0 \\ \hline 0 & B(g) \end{array} \right)$$

where $A(g)$ and $B(g)$ are the matrices of the action of G restricted to U and W, respectively.

Returning to the defining representation of $\mathcal{S}_3$, we see that

$$V = \mathbf{C}[\mathbf{1}, \mathbf{2}, \mathbf{3}] = \mathbf{C}[\mathbf{1} + \mathbf{2} + \mathbf{3}] \oplus \mathbf{C}[\mathbf{2}, \mathbf{3}]$$

as vector spaces. But, while $\mathbf{C}[\mathbf{1} + \mathbf{2} + \mathbf{3}]$ is an $\mathcal{S}_3$-submodule, $\mathbf{C}[\mathbf{2}, \mathbf{3}]$ is not (e.g., $(1, 2)\mathbf{2} = \mathbf{1} \notin \mathbf{C}[\mathbf{2}, \mathbf{3}]$). So we need to find a complement for $\mathbf{C}[\mathbf{1} + \mathbf{2} + \mathbf{3}]$—i.e., a *submodule* U such that

$$\mathbf{C}[\mathbf{1}, \mathbf{2}, \mathbf{3}] = \mathbf{C}[\mathbf{1} + \mathbf{2} + \mathbf{3}] \oplus U$$

To find a complement, we introduce an inner product on $\mathbf{C}[\mathbf{1}, \mathbf{2}, \mathbf{3}]$. Given any two vectors $\mathbf{i}, \mathbf{j}$ in the basis $\{\mathbf{1}, \mathbf{2}, \mathbf{3}\}$, let their inner product be

$$\langle \mathbf{i}, \mathbf{j} \rangle = \delta_{i,j} \tag{1.5}$$

where $\delta_{i,j}$ is the Kronecker delta. Now we extend by linearity in the first variable and conjugate linearity in the second to obtain an inner product on the whole vector space. Equivalently, we could have started out by defining the product of any two given vectors $\mathbf{v} = a\mathbf{1} + b\mathbf{2} + c\mathbf{3}$, $\mathbf{w} = x\mathbf{1} + y\mathbf{2} + z\mathbf{3}$ as

$$\langle \mathbf{v}, \mathbf{w} \rangle = a\bar{x} + b\bar{y} + c\bar{z}$$

with the bar denoting complex conjugation. The reader can check that this definition does indeed satisfy all the axioms for an inner product. It also enjoys the property that it is *invariant* under the action of G:

$$\langle g\mathbf{v}, g\mathbf{w} \rangle = \langle \mathbf{v}, \mathbf{w} \rangle \qquad \text{for all } g \in G \text{ and } \mathbf{v}, \mathbf{w} \in V \tag{1.6}$$

To check invariance on V, it suffices to verify (1.6) for elements of a basis. But if $\pi \in \mathcal{S}_3$, then

$$\langle \pi\mathbf{i}, \pi\mathbf{j} \rangle = \delta_{\pi(i), \pi(j)} = \delta_{i,j} = \langle \mathbf{i}, \mathbf{j} \rangle$$

where the middle equality holds since π is a bijection.

Now, given any inner product on a vector space V and a subspace W, we can form the *orthogonal complement*:

$$W^{\perp} = \{\mathbf{v} \in V \mid \langle \mathbf{v}, \mathbf{w} \rangle = 0 \text{ for all } \mathbf{w} \in W\}$$

It is always true that $V = W \oplus W^{\perp}$. When $W \le V$ and the inner product is G-invariant, we can say more.

Proposition 1.5.2 *Let V be a G-module, W a submodule, and $\langle \cdot, \cdot \rangle$ an inner product invariant under the action of G. Then $W^{\perp}$ is also a G-submodule.*

Proof. We must show that for all $g \in G$ and $\mathbf{u} \in W^{\perp}$ we have $g\mathbf{u} \in W^{\perp}$. Take any $\mathbf{w} \in W$; then

$$\begin{aligned}
\langle g\mathbf{u}, \mathbf{w} \rangle &= \langle g^{-1}g\mathbf{u}, g^{-1}\mathbf{w} \rangle && \text{(since } \langle \cdot, \cdot \rangle \text{ is invariant)} \\
&= \langle \mathbf{u}, g^{-1}\mathbf{w} \rangle && \text{(properties of group action)} \\
&= 0 && (\mathbf{u} \in W^{\perp}, \text{ and } g^{-1}\mathbf{w} \in W \\
& && \text{since } W \text{ is a submodule)}
\end{aligned}$$

Thus $W^{\perp}$ is closed under the action of G. ∎

Applying this to our running example, we see that

$$\begin{aligned}
\mathbf{C}[\mathbf{1} + \mathbf{2} + \mathbf{3}]^{\perp} &= \{\mathbf{v} = a\mathbf{1} + b\mathbf{2} + c\mathbf{3} \mid \langle \mathbf{v}, \mathbf{1} + \mathbf{2} + \mathbf{3} \rangle = 0\} \\
&= \{\mathbf{v} = a\mathbf{1} + b\mathbf{2} + c\mathbf{3} \mid a + b + c = 0\}
\end{aligned}$$

To compute the matrices of the direct sum, we choose the bases $\{\mathbf{1} + \mathbf{2} + \mathbf{3}\}$ for $\mathbf{C}[\mathbf{1} + \mathbf{2} + \mathbf{3}]$, and $\{\mathbf{2} - \mathbf{1}, \ \mathbf{3} - \mathbf{1}\}$ for $\mathbf{C}[\mathbf{1} + \mathbf{2} + \mathbf{3}]^{\perp}$. This produces the matrices

$$X(e) = \begin{pmatrix} 1 & 0 & 0 \\ 0 & 1 & 0 \\ 0 & 0 & 1 \end{pmatrix}, \qquad X((1,2)) = \begin{pmatrix} 1 & 0 & 0 \\ 0 & -1 & -1 \\ 0 & 0 & 1 \end{pmatrix}$$

$$X((1,3)) = \begin{pmatrix} 1 & 0 & 0 \\ 0 & 1 & 0 \\ 0 & -1 & -1 \end{pmatrix}, \qquad X((2,3)) = \begin{pmatrix} 1 & 0 & 0 \\ 0 & 0 & 1 \\ 0 & 1 & 0 \end{pmatrix}$$

$$X((1,2,3)) = \begin{pmatrix} 1 & 0 & 0 \\ 0 & -1 & -1 \\ 0 & 1 & 0 \end{pmatrix}, \qquad X((1,3,2)) = \begin{pmatrix} 1 & 0 & 0 \\ 0 & 0 & 1 \\ 0 & -1 & -1 \end{pmatrix}$$

These are indeed all direct matrix sums of the form

$$X(g) = \left(\begin{array}{c|cc} A(g) & 0 & 0 \\ \hline 0 & & \\ 0 & & B(g) \end{array} \right)$$

Of course, $A(g)$ is irreducible (being of degree 1) and we will see in Section 1.9 that $B(g)$ is also. Thus we have decomposed the defining representation of $\mathcal{S}_3$ into its irreducible parts. The content of Maschke's theorem is that this can be done for any finite group.

Theorem 1.5.3 (Maschke's Theorem) *Let G be a finite group and let V be a nonzero G-module. Then*

$$V = W^{(1)} \oplus W^{(2)} \oplus \cdots \oplus W^{(k)}$$

where each $W^{(i)}$ is an irreducible G-submodule of V.

Proof. We will induct on $d = \dim V$. If $d = 1$, then V itself is irreducible and we are done ($k = 1$ and $W^{(1)} = V$). Now suppose that $d > 1$. If V is irreducible, then we are finished as before. If not, then V has a nontrivial G-submodule, W. We will try to construct a submodule complement for W as we did in the preceding example.

Pick any basis $\mathcal{B} = \{\mathbf{v}_1, \mathbf{v}_2, \ldots, \mathbf{v}_d\}$ for V. Consider the unique inner product that satisfies

$$\langle \mathbf{v}_i, \mathbf{v}_j \rangle = \delta_{i,j}$$

for elements of $\mathcal{B}$. This product may not be G-invariant, but we can come up with another one that is. For any $\mathbf{v}, \mathbf{w} \in V$ we let

$$\langle \mathbf{v}, \mathbf{w} \rangle' = \sum_{g \in G} \langle g\mathbf{v}, g\mathbf{w} \rangle$$

We leave it to the reader to verify that $\langle \cdot, \cdot \rangle'$ satisfies the definition of an inner product. To show that it is G-invariant, we wish to prove

$$\langle h\mathbf{v}, h\mathbf{w} \rangle' = \langle \mathbf{v}, \mathbf{w} \rangle'$$

for all $h \in G$ and $\mathbf{v}, \mathbf{w} \in V$. But

$$
\begin{aligned}
\langle h\mathbf{v}, h\mathbf{w} \rangle' &= \sum_{g \in G} \langle gh\mathbf{v}, gh\mathbf{w} \rangle && \text{(definition of } \langle \cdot, \cdot \rangle' \text{)} \\
&= \sum_{f \in G} \langle f\mathbf{v}, f\mathbf{w} \rangle && \text{(as } g \text{ varies over } G \text{, so does } f = gh \text{)} \\
&= \langle \mathbf{v}, \mathbf{w} \rangle' && \text{(definition of } \langle \cdot, \cdot \rangle' \text{)}
\end{aligned}
$$

as desired.

If we let

$$W^{\perp} = \{\mathbf{v} \in V \mid \langle \mathbf{v}, \mathbf{w} \rangle' = 0\}$$

then by Proposition 1.5.2 we have that $W^{\perp}$ is a G-submodule of V with

$$V = W \oplus W^{\perp}$$

Now we can apply induction to W and $W^{\perp}$ to write each as a direct sum of irreducibles. Putting these two decompositions together, we see that V has the desired form. ∎

As a corollary, we have the matrix version of Maschke's theorem. Here and in the future, we often drop the horizontal and vertical lines indicating block matrices. Our convention of using lowercase letters for elements and uppercase ones for matrices should avoid any confusion.

Corollary 1.5.4 *Let G be a finite group and let X be a matrix representation of G of dimension $d > 0$. Then there is a fixed matrix T such that every matrix $X(g)$, $g \in G$, has the form*

$$TX(g)T^{-1} = \begin{pmatrix} X^{(1)}(g) & 0 & \cdots & 0 \\ 0 & X^{(2)}(g) & \cdots & 0 \\ \vdots & \vdots & \ddots & \vdots \\ 0 & 0 & \cdots & X^{(k)}(g) \end{pmatrix}$$

where each $X^{(i)}$ is an irreducible matrix representation of G.

Proof. Let $V = \mathbf{C}^d$ with the action

$$g\mathbf{v} = X(g)\mathbf{v}$$

for all $g \in G$ and $\mathbf{v} \in V$. By Maschke's theorem,

$$V = W^{(1)} \oplus W^{(2)} \oplus \cdots \oplus W^{(k)}$$

each $W^{(i)}$ being irreducible of dimension, say, d_i. Take a basis $\mathcal{B}$ for V such that the first d_1 vectors are a basis for $W^{(1)}$, the next d_2 are a basis for $W^{(2)}$, etc. The matrix T that transforms the standard basis for $\mathbf{C}^d$ into $\mathcal{B}$ now does the trick, since conjugating by T just expresses each $X(g)$ in the new basis $\mathcal{B}$. ∎

Representations that decompose so nicely have a name.

Definition 1.5.5 A representation is *completely reducible* if it can be written as a direct sum of irreducibles. ∎

So Maschke's theorem could be restated:

> *Every representation of a finite group having positive dimension is completely reducible.*

We are working under the nicest possible assumptions, namely, that all our groups are finite and all our vector spaces are over $\mathbf{C}$. We will, however, occasionally attempt to indicate more general results. Maschke's theorem remains true if $\mathbf{C}$ is replaced by any field whose characteristic is either zero or prime to $|G|$. For a proof in this setting, the reader can consult Ledermann [Led 77, pages 21–23].

However, we can *not* drop the finiteness assumption on G, as the following example shows. Let $\mathbf{R}^+$ be the positive real numbers, which are a group under multiplication. It is not hard to see that letting

$$X(r) = \begin{pmatrix} 1 & \log r \\ 0 & 1 \end{pmatrix}$$

for all $r \in \mathbf{R}^+$ defines a representation. The subspace

$$W = \left\{ \begin{pmatrix} c \\ 0 \end{pmatrix} \mid c \in \mathbf{C} \right\} \subset \mathbf{C}^2$$

is invariant under the action of G. Thus if X is completely reducible, then $\mathbf{C}^2$ must decompose as the direct sum of W and another one-dimensional submodule. By the matrix version of Maschke's theorem, there exists a fixed matrix T such that

$$TX(r)T^{-1} = \left(\begin{array}{cc} x(r) & 0 \\ 0 & y(r) \end{array} \right)$$

for all $r \in \mathbf{R}^+$. Thus $x(r)$ and $y(r)$ must be the eigenvalues of $X(r)$, which are both 1. But then

$$\begin{aligned} X(r) &= T^{-1} \left(\begin{array}{cc} 1 & 0 \\ 0 & 1 \end{array} \right) T \\ &= \left(\begin{array}{cc} 1 & 0 \\ 0 & 1 \end{array} \right) \end{aligned}$$

for all $r \in \mathbf{R}^+$, which is absurd.

1.6 G-homomorphisms and Schur's Lemma

We can learn more about an object in mathematics (e.g., vector spaces, groups, topological spaces) by studying functions that preserve the structure of that object (e.g., linear transformations, homomorphisms, continuous maps). For a G-module, the corresponding function is called a G-homomorphism.

Definition 1.6.1 Let V and W be G-modules. Then a *G-homomorphism* (or simply a *homomorphism*) is a linear transformation $\theta : V \rightarrow W$ such that

$$\theta(g\mathbf{v}) = g\theta(\mathbf{v})$$

for all $g \in G$ and $\mathbf{v} \in V$. We also say that θ *preserves* or *respects* the action of G. ∎

We can translate this into the language of matrices by taking bases $\mathcal{B}$ and $\mathcal{C}$ for V and W, respectively. Let $X(g)$ and $Y(g)$ be the corresponding matrix representations. Also take T to be the matrix of θ in the two bases $\mathcal{B}$ and $\mathcal{C}$. Then the G-homomorphism property becomes

$$TX(g)\mathbf{v} = Y(g)T\mathbf{v}$$

for every column vector $\mathbf{v}$ and $g \in G$. But since this holds for all $\mathbf{v}$, we must have

$$TX(g) = Y(g)T \qquad \text{for all } g \in G \tag{1.7}$$

Thus having a G-homomorphism θ is equivalent to the existence of a matrix T such that (1.7) holds. We will often write this condition simply as $TX = YT$.

As an example, let $G = \mathcal{S}_n$, $V = \mathbf{C}[\mathbf{v}]$ with the trivial action of $\mathcal{S}_n$, and let $W = \mathbf{C}[\mathbf{1}, \mathbf{2}, \ldots, \mathbf{n}]$ with the defining action of $\mathcal{S}_n$. Define a transformation $\theta : V \to W$ by

$$\theta(\mathbf{v}) = \mathbf{1} + \mathbf{2} + \cdots + \mathbf{n}$$

and linear extension; i.e.,

$$\theta(c\mathbf{v}) = c(\mathbf{1} + \mathbf{2} + \cdots + \mathbf{n})$$

for all $c \in \mathbf{C}$. To check that θ is a G-homomorphism, it suffices to check that the action of G is preserved on a basis of V. (Why?) But, for all $\pi \in \mathcal{S}_n$,

$$\theta(\pi\mathbf{v}) = \theta(\mathbf{v}) = \sum_{i=1}^{n} \mathbf{i} = \pi \sum_{i=1}^{n} \mathbf{i} = \pi\theta(\mathbf{v})$$

In a similar vein, let G be an arbitrary group acting trivially on $V = \mathbf{C}[\mathbf{v}]$, and let $W = \mathbf{C}[\mathbf{G}]$ is the group algebra. Now we have the G-homomorphism $\theta : V \to W$ given by extending

$$\theta(\mathbf{v}) = \sum_{g \in G} \mathbf{g}$$

linearly.

If $G = \mathcal{S}_n$, we can also let G act on $V = \mathbf{C}[\mathbf{v}]$ by using the sign representation:

$$\pi\mathbf{u} = \operatorname{sgn}(\pi)\mathbf{u}$$

for all $\pi \in \mathcal{S}_n$ and $\mathbf{u} \in V$. Keeping the usual action on the group algebra, the reader can verify that

$$\eta(\mathbf{v}) = \sum_{\pi \in \mathcal{S}_n} \operatorname{sgn}(\pi)\boldsymbol{\pi}$$

extends to a G-homomorphism from V to W.

It is clearly important to know when two representations of a group are different and when they are not (even though there may be some cosmetic differences). For example, two matrix representations that differ only by a basis change are really the same. The concept of G-equivalence captures this idea.

Definition 1.6.2 Let V and W be modules for a group G. A *G-isomorphism* is a G-homomorphism $\theta : V \to W$ that is bijective. In this case we say that V and W are *G-isomorphic*, or *G-equivalent*, written $V \cong W$. Otherwise we say that V and W are *G-inequivalent*. ∎

As usual, we drop the G when the group is implied by context.

In matrix terms, θ being a bijection translates into the corresponding matrix T being invertible. Thus from equation (1.7) we see that matrix representations X and Y of a group G are equivalent if and only if there exists a fixed matrix T such that

$$Y(g) = TX(g)T^{-1}$$

for all $g \in G$. This is the change-of-basis criterion that we were talking about earlier.

Example 1.6.3 We are now in a position to explain why the coset representation of $\mathcal{S}_3$ at the end of Example 1.3.5 is the same as the defining representation. Recall that we had taken the subgroup $H = \{\epsilon, (2,3)\} \subset \mathcal{S}_3$ giving rise to the coset representation module $\mathbf{C}[\mathcal{H}]$, where

$$\mathcal{H} = \{H , (1,2)H , (1,3)H\}$$

Given any set A, let $\mathcal{S}_A$ be the symmetric group on A, i.e., the set of all permutations of A. Now the subgroup H can be expressed as an (internal) direct product

$$H = \{(1)(2)(3), (1)(2,3)\} = \{(1)\} \times \{(2)(3), (2,3)\} = \mathcal{S}_{\{1\}} \times \mathcal{S}_{\{2,3\}} \quad (1.8)$$

A convenient device for displaying such product subgroups of $\mathcal{S}_n$ is the tabloid. Let $\lambda = (\lambda_1, \lambda_2, \ldots, \lambda_l)$ be a partition, as discussed in Section 1.1. A *Young tabloid of shape* λ is an array with l rows such that row i contains λ_i integers and the order of entries in a row does not matter. To show that each row can be shuffled arbitrarily, we put horizontal lines between the rows. For example, if $\lambda = (4, 2, 1)$, then some of the possible Young tabloids are

$$
\begin{array}{c}
\overline{\begin{array}{cccc} 3 & 1 & 4 & 1 \end{array}} \\
\overline{\begin{array}{cc} 5 & 9 \end{array}} \\
\overline{\begin{array}{c} 2 \end{array}}
\end{array}
\quad = \quad
\begin{array}{c}
\overline{\begin{array}{cccc} 3 & 1 & 1 & 4 \end{array}} \\
\overline{\begin{array}{cc} 9 & 5 \end{array}} \\
\overline{\begin{array}{c} 2 \end{array}}
\end{array}
\quad \neq \quad
\begin{array}{c}
\overline{\begin{array}{cccc} 9 & 5 & 3 & 4 \end{array}} \\
\overline{\begin{array}{cc} 2 & 1 \end{array}} \\
\overline{\begin{array}{c} 1 \end{array}}
\end{array}
$$

Equation (1.8) says that H consists of all permutations in $\mathcal{S}_3$ that permute the elements of the set $\{1\}$ among themselves (giving only the permutation (1)) and permute the elements of $\{2,3\}$ among themselves (giving $(2)(3)$ and $(2,3)$). This is modeled by the tabloid

$$
\begin{array}{c}
\overline{\begin{array}{cc} 2 & 3 \end{array}} \\
\overline{\begin{array}{c} 1 \end{array}}
\end{array}
$$

since the order of 2 and 3 is immaterial but 1 must remain fixed. The complete set of tabloids of shape $\lambda = (2,1)$ whose entries are exactly 1, 2, 3 is

$$S = \left\{
\begin{array}{c}
\overline{\begin{array}{cc} 2 & 3 \end{array}} \\
\overline{\begin{array}{c} 1 \end{array}}
\end{array}
,\quad
\begin{array}{c}
\overline{\begin{array}{cc} 1 & 3 \end{array}} \\
\overline{\begin{array}{c} 2 \end{array}}
\end{array}
,\quad
\begin{array}{c}
\overline{\begin{array}{cc} 1 & 2 \end{array}} \\
\overline{\begin{array}{c} 3 \end{array}}
\end{array}
\right\}$$

Furthermore, there is an action of any $\pi \in \mathcal{S}_3$ on S given by

$$
\pi \begin{array}{c}
\overline{\begin{array}{cc} i & j \end{array}} \\
\overline{\begin{array}{c} k \end{array}}
\end{array}
\quad = \quad
\begin{array}{c}
\overline{\begin{array}{cc} \pi(i) & \pi(j) \end{array}} \\
\overline{\begin{array}{c} \pi(k) \end{array}}
\end{array}
$$

Thus it makes sense to consider the map θ that sends

$$
\mathbf{H} \quad \overset{\theta}{\to} \quad
\begin{array}{c}
\overline{\begin{array}{cc} 2 & 3 \end{array}} \\
\overline{\begin{array}{c} 1 \end{array}}
\end{array}
$$

$$(1,2)\mathbf{H} \quad \xrightarrow{\theta} \quad (1,2)\frac{\boxed{2\ 3}}{\boxed{1}} = \frac{\boxed{1\ 3}}{\boxed{2}}$$

$$(1,3)\mathbf{H} \quad \xrightarrow{\theta} \quad (1,3)\frac{\boxed{2\ 3}}{\boxed{1}} = \frac{\boxed{1\ 2}}{\boxed{3}}.$$

By linear extension, θ becomes a vector space isomorphism from $\mathbf{C}[\mathcal{H}]$ to $\mathbf{C}[\mathbf{S}]$. In fact, we claim it is also a G-isomorphism. To verify this, we can check that the action of each $\pi \in \mathcal{S}_3$ is preserved on each basis vector in H. For example, if $\pi = (1,2)$ and $\mathbf{H} \in \mathcal{H}$, then

$$\theta((1,2)\mathbf{H}) = \theta((1,2)\mathbf{H}) = \frac{\boxed{1\ 3}}{\boxed{2}} = (1,2)\frac{\boxed{2\ 3}}{\boxed{1}} = (1,2)\theta(\mathbf{H})$$

Thus

$$\mathbf{C}[\mathcal{H}] \cong \mathbf{C}[\mathbf{S}] \tag{1.9}$$

Another fact about the tabloids in our set S is that they are completely determined by the element placed in the second row. So we have a natural map, η, between the basis $\{\mathbf{1,2,3}\}$ for the defining representation and $\mathbf{S}$, namely,

$$1 \quad \xrightarrow{\eta} \quad \frac{\boxed{2\ 3}}{\boxed{1}}$$

$$2 \quad \xrightarrow{\eta} \quad \frac{\boxed{1\ 3}}{\boxed{2}}$$

$$3 \quad \xrightarrow{\eta} \quad \frac{\boxed{1\ 2}}{\boxed{3}}$$

Now η extends by linearity to a G-isomorphism from $\mathbf{C}[\mathbf{1,2,3}]$ to $\mathbf{C}[\mathbf{S}]$. This, in combination with equation (1.9), shows that $\mathbf{C}[\mathcal{H}]$ and $\mathbf{C}[\mathbf{1,2,3}]$ are indeed equivalent.

The reader may feel that we have taken a long and windy route to get to the final $\mathcal{S}_3$-isomorphism. However, the use of Young tabloids extends far beyond this example. In fact, we use them to construct all the irreducible representations of $\mathcal{S}_n$ in Chapter 2. ∎

We now return to the general exposition. Two sets usually associated with any map of vector spaces $\theta : V \rightarrow W$ are its *kernel*,

$$\ker\theta = \{\mathbf{v} \in V \mid \theta(\mathbf{v}) = \mathbf{0}\}$$

where $\mathbf{0}$ is the zero vector, and its *image*,

$$\operatorname{im}\theta = \{\mathbf{w} \in W \mid \mathbf{w} = \theta(\mathbf{v}) \text{ for some } \mathbf{v} \in V\}$$

When θ is a G-homomorphism, the kernel and image have nice structure.

Proposition 1.6.4 *Let $\theta : V \to W$ be a G-homomorphism. Then*

1. *$\ker \theta$ is a G-submodule of V, and*

2. *$\operatorname{im} \theta$ is a G-submodule of W.*

Proof. We prove only the first assertion, leaving the second one for the reader. It is known from the theory of vector spaces that $\ker \theta$ is a subspace of V since θ is linear. So we only need to show closure under the action of G. But if $\mathbf{v} \in \ker \theta$, then for any $g \in G$:

$$
\begin{aligned}
\theta(g\mathbf{v}) \;&=\; g\theta(\mathbf{v}) \quad (\theta \text{ is a } G\text{-homomorphism}) \\
&=\; g\mathbf{0} \qquad (\mathbf{v} \in \ker \theta) \\
&=\; \mathbf{0}
\end{aligned}
$$

and so $g\mathbf{v} \in \ker \theta$, as desired. ∎

It is now an easy matter to prove Schur's lemma, which characterizes G-homomorphisms of irreducible modules. This result plays a crucial role when we discuss the commutant algebra in the next section.

Theorem 1.6.5 (Schur's Lemma) *Let V and W be two irreducible G-modules. If $\theta : V \to W$ is a G-homomorphism, then either*

1. *θ is a G-isomorphism, or*

2. *θ is the zero map.*

Proof. Since V is irreducible and $\ker \theta$ is a submodule (by the previous proposition), we must have either $\ker \theta = \{\mathbf{0}\}$ or $\ker \theta = V$. Similarly the irreducibility of W implies that $\operatorname{im} \theta = \{\mathbf{0}\}$ or W. If $\ker \theta = V$ or $\operatorname{im} \theta = \{\mathbf{0}\}$, then θ must be the zero map. On the other hand, if $\ker \theta = \{\mathbf{0}\}$ and $\operatorname{im} \theta = W$, then we have an isomorphism. ∎

It is interesting to note that Schur's lemma continues to be valid over arbitrary fields and for infinite groups. In fact, the proof we just gave still works. The matrix version is also true in this more general setting

Corollary 1.6.6 *Let X and Y be two irreducible matrix representations of G. If T is any matrix such that $TX(g) = Y(g)T$ for all $g \in G$, then either*

1. *T is invertible, or*

2. *T is the zero matrix.* ∎

We also have an analog of Schur's lemma in the case where the range module is not irreducible. This result is conveniently expressed in terms of the vector space $\operatorname{Hom}(V, W)$ of all G-homomorphisms from V to W.

Corollary 1.6.7 *Let V and W be two G-modules with V being irreducible. Then $\dim \operatorname{Hom}(V, W) = 0$ if and only if W contains no submodule isomorphic to V.* ∎

When the field is $\mathbf{C}$, however, we can say more. Suppose that T is a matrix such that

$$TX(g) = X(g)T \qquad (1.10)$$

for all $g \in G$. It follows that

$$(T - cI)X = X(T - cI)$$

where I is the appropriate identity matrix and $c \in \mathbf{C}$ is any scalar. Now $\mathbf{C}$ is algebraically closed, so we can take c to be an eigenvalue of T. Thus $T - cI$ satisfies the hypothesis of Corollary 1.6.6 (with $X = Y$) and is not invertible by the choice of c. Our only alternative is that $T - cI = 0$. We have proved:

Corollary 1.6.8 *Let X be an irreducible matrix representation of G over the complex numbers. Then the only matrices T that commute with $X(g)$ for all $g \in G$ are those of the form $T = cI$—i.e., scalar multiples of the identity matrix.* ■

1.7 Commutant and Endomorphism Algebras

Corollary 1.6.8 suggests that the set of matrices that commute with those of a given representation are important. This corresponds in the module setting to the set of G-homomorphisms from a G-module to itself. We characterize these sets in this section. Extending these ideas to homomorphisms between different G-modules leads to a useful generalization of Corollary 1.6.7 (see Corollary 1.7.10).

Definition 1.7.1 Given a matrix representation $X : G \to GL_d$, the corresponding *commutant algebra* is

$$\operatorname{Com} X = \{T \in \operatorname{Mat}_d \mid TX(g) = X(g)T \text{ for all } g \in G\}$$

where Mat_d is the set of all $d \times d$ matrices with entries in $\mathbf{C}$. Given a G-module V, the corresponding *endomorphism algebra* is

$$\operatorname{End} V = \{\theta : V \to V \mid \theta \text{ is a } G\text{-homomorphism}\} \ \blacksquare$$

It is easy to check that both the commutant and endomorphism algebras do satisfy the axioms for an algebra. The reader can also verify that if V is a G-module and X is a corresponding matrix representation, then $\operatorname{End} V$ and $\operatorname{Com} X$ are isomorphic as algebras. Merely take the basis $\mathcal{B}$ that produced X and use the map that sends each $\theta \in \operatorname{End} V$ to the matrix T of θ in the basis $\mathcal{B}$. Let us compute $\operatorname{Com} X$ for various representations X.

Example 1.7.2 Suppose that X is a matrix representation such that

$$X = \begin{pmatrix} X^{(1)} & 0 \\ 0 & X^{(2)} \end{pmatrix} = X^{(1)} \oplus X^{(2)}$$

where $X^{(1)}, X^{(2)}$ are inequivalent and irreducible of degrees d_1, d_2, respectively. What does $\text{Com}\, X$ look like?

Suppose that

$$T = \begin{pmatrix} T_{1,1} & T_{1,2} \\ T_{2,1} & T_{2,2} \end{pmatrix}$$

is a matrix partitioned in the same way as X. If $TX = XT$, then we can multiply out each side to obtain

$$\begin{pmatrix} T_{1,1}X^{(1)} & T_{1,2}X^{(2)} \\ T_{2,1}X^{(1)} & T_{2,2}X^{(2)} \end{pmatrix} = \begin{pmatrix} X^{(1)}T_{1,1} & X^{(1)}T_{1,2} \\ X^{(2)}T_{2,1} & X^{(2)}T_{2,2} \end{pmatrix}$$

Equating corresponding blocks we get

$$\begin{aligned} T_{1,1}X^{(1)} &= X^{(1)}T_{1,1} \\ T_{1,2}X^{(2)} &= X^{(1)}T_{1,2} \\ T_{2,1}X^{(1)} &= X^{(2)}T_{2,1} \\ T_{2,2}X^{(2)} &= X^{(2)}T_{2,2} \end{aligned}$$

Using Corollaries 1.6.6 and 1.6.8 along with the fact that $X^{(1)}$ and $X^{(2)}$ are inequivalent, these equations can be solved to yield

$$T_{1,1} = c_1 I_{d_1}, \qquad T_{1,2} = T_{2,1} = 0, \qquad T_{2,2} = c_2 I_{d_2}$$

where $c_1, c_2 \in \mathbf{C}$ and I_{d_1}, I_{d_2} are identity matrices of degrees d_1, d_2. Thus

$$T = \begin{pmatrix} c_1 I_{d_1} & 0 \\ 0 & c_2 I_{d_2} \end{pmatrix}$$

We have shown that when $X = X^{(1)} \oplus X^{(2)}$ with $X^{(1)} \not\cong X^{(2)}$ and irreducible, then

$$\text{Com}\, X = \{ c_1 I_{d_1} \oplus c_2 I_{d_2} \mid c_1, c_2 \in \mathbf{C} \}$$

where $d_1 = \deg X^{(1)}$, $d_2 = \deg X^{(2)}$. ∎

In general, if $X = \oplus_{i=1}^{k} X^{(i)}$, where the $X^{(i)}$ are pairwise inequivalent irreducibles, then a similar argument proves that

$$\text{Com}\, X = \{ \oplus_{i=1}^{k} c_i I_{d_i} \mid c_i \in \mathbf{C} \}$$

where $d_i = \deg X^{(i)}$. Notice that the degree of X is $\sum_{i=1}^{k} d_i$. Note, also, that the dimension of $\text{Com}\, X$ (as a vector space) is just k. This is because there are k scalars c_i that can vary, whereas the identity matrices are fixed.

Next we deal with the case of sums of equivalent representations. A convenient notation is

$$mX = \overbrace{X \oplus X \oplus \cdots \oplus X}^{m}$$

where the nonnegative integer m is called the *multiplicity of X*.

Example 1.7.3 Suppose that

$$X = \begin{pmatrix} X^{(1)} & 0 \\ 0 & X^{(1)} \end{pmatrix} = 2X^{(1)}$$

where $X^{(1)}$ is irreducible of degree d. Take T partitioned as before. Doing the multiplication in $TX = XT$ and equating blocks now yields four equations, all of the form

$$T_{i,j}X^{(1)} = X^{(1)}T_{i,j}$$

for all $i, j = 1, 2$. Corollaries 1.6.6 and 1.6.8 come into play again to reveal that, for all i and j,

$$T_{i,j} = c_{i,j}I_d$$

where $c_{i,j} \in \mathbf{C}$. Thus

$$\text{Com}\, X = \left\{ \begin{pmatrix} c_{1,1}I_d & c_{1,2}I_d \\ c_{2,1}I_d & c_{2,2}I_d \end{pmatrix} \,\Big|\, c_{i,j} \in \mathbf{C} \text{ for all } i, j \right\} \tag{1.11}$$

is the commutant algebra in this case. ∎

The matrices in $\text{Com}\, 2X^{(1)}$ have a name.

Definition 1.7.4 Let $X = (x_{i,j})$ and Y be matrices. Then their *tensor product* is the block matrix

$$X \otimes Y = (x_{i,j}Y) = \begin{pmatrix} x_{1,1}Y & x_{1,2}Y & \cdots \\ x_{2,1}Y & x_{2,2}Y & \cdots \\ \vdots & \vdots & \ddots \end{pmatrix} \quad \blacksquare$$

Thus we could write the elements of (1.11) as

$$T = \begin{pmatrix} c_{1,1} & c_{1,2} \\ c_{2,1} & c_{2,2} \end{pmatrix} \otimes I_d$$

and so

$$\text{Com}\, X = \{M_2 \otimes I_d \mid M_2 \in \text{Mat}_2\}$$

If we take $X = mX^{(1)}$, then

$$\text{Com}\, X = \{M_m \otimes I_d \mid M_m \in \text{Mat}_m\}$$

where d is the degree of $X^{(1)}$. Computing degrees and dimensions, we obtain

$$\deg X = \deg mX^{(1)} = m \deg X^{(1)} = md$$

and

$$\dim(\text{Com}\, X) = \dim\{M_m \mid M_m \in \text{Mat}_m\} = m^2$$

Finally, we are led to consider the most general case:

$$X = m_1 X^{(1)} \oplus m_2 X^{(2)} \oplus \cdots \oplus m_k X^{(k)} \tag{1.12}$$

where the $X^{(i)}$ are pairwise inequivalent irreducibles with $\deg X^{(i)} = d_i$. The degree of X is given by

$$\deg X = \sum_{i=1}^{k} \deg(m_i X^{(i)}) = m_1 d_1 + m_2 d_2 + \cdots + m_k d_k$$

The reader should have no trouble combining Examples 1.7.2 and 1.7.3 to obtain

$$\operatorname{Com} X = \{\oplus_{i=1}^{k}(M_{m_i} \otimes I_{d_i}) \mid M_{m_i} \in \operatorname{Mat}_{m_i} \text{ for all } i\} \qquad (1.13)$$

of dimension

$$\dim(\operatorname{Com} X) = \dim\{\oplus_{i=1}^{k} M_{m_i} \mid M_{m_i} \in \operatorname{Mat}_{m_i}\} = m_1^2 + m_2^2 + \cdots + m_k^2$$

Before continuing our investigation of the commutant algebra, we should briefly mention the abstract vector space analog of the tensor product.

Definition 1.7.5 Given vector spaces V and W, then their *tensor product* is the set

$$V \otimes W = \{\sum_{i,j} c_{i,j} \mathbf{v}_i \otimes \mathbf{w}_j \mid c_{i,j} \in \mathbf{C}, \mathbf{v}_i \in V, \mathbf{w}_j \in W\}$$

subject to the relations

$$
\begin{aligned}
(c_1 \mathbf{v}_1 + c_2 \mathbf{v}_2) \otimes \mathbf{w} &= c_1(\mathbf{v}_1 \otimes \mathbf{w}) + c_2(\mathbf{v}_2 \otimes \mathbf{w}) \text{ and} \\
\mathbf{v} \otimes (d_1 \mathbf{w}_1 + d_2 \mathbf{w}_2) &= d_1(\mathbf{v} \otimes \mathbf{w}_1) + d_2(\mathbf{v} \otimes \mathbf{w}_2) \ \blacksquare
\end{aligned}
$$

It is easy to see that $V \otimes W$ is also a vector space. In fact, if $\mathcal{B} = \{\mathbf{v}_1, \mathbf{v}_2, \ldots, \mathbf{v}_d\}$ and $\mathcal{C} = \{\mathbf{w}_1, \mathbf{w}_2, \ldots, \mathbf{w}_f\}$ are bases for V and W, respectively, then the set

$$\{\mathbf{v}_i \otimes \mathbf{w}_j \mid 1 \le i \le d, 1 \le j \le f\}$$

is a basis for $V \otimes W$. This gives the connection with the definition of matrix tensor products: the algebra Mat_d has as basis the set

$$\mathcal{B} = \{E_{i,j} \mid 1 \le i, j \le d\}$$

where $E_{i,j}$ is the matrix of zeros with exactly one 1 in position (i, j). So if $X = (x_{i,j}) \in \operatorname{Mat}_d$ and $Y = (y_{k,l}) \in \operatorname{Mat}_f$, then, by the fact that $\otimes$ is linear,

$$
\begin{aligned}
X \otimes Y &= \left(\sum_{i,j=1}^{d} x_{i,j} E_{i,j}\right) \otimes \left(\sum_{k,l=1}^{f} y_{k,l} E_{k,l}\right) \\
&= \sum_{i,j=1}^{d} \sum_{k,l=1}^{f} x_{i,j} y_{k,l} (E_{i,j} \otimes E_{k,l}) \qquad (1.14)
\end{aligned}
$$

But if $E_{i,j} \otimes E_{k,l}$ represents the $(k,l)^{\text{th}}$ position of the $(i,j)^{\text{th}}$ block of a matrix, then equation (1.14) says that the corresponding entry for $X \otimes Y$ should be $x_{i,j} y_{k,l}$, agreeing with the matrix definition.

We return from our brief detour to consider the center of $\text{Com}\, X$. The *center* of an algebra A is

$$Z_A = \{a \in A \mid ab = ba \text{ for all } b \in A\}$$

First we will compute the center of a matrix algebra. This result should be very reminiscent of Corollary 1.6.8 to Schur's lemma.

Proposition 1.7.6 *The center of* Mat_d *is*

$$Z_{\text{Mat}_d} = \{cI_d \mid c \in \mathbf{C}\}$$

Proof. Suppose that $C \in Z_{\text{Mat}_d}$. Then, in particular,

$$CE_{i,i} = E_{i,i}C \qquad (1.15)$$

for all i. But $CE_{i,i}$ (respectively, $E_{i,i}C$) is all zeros except for the ith column (respectively, row), which is the same as C's. Thus (1.15) implies that all off-diagonal elements of C must be 0. Similarly, if $i \neq j$:

$$C(E_{i,j} + E_{j,i}) = (E_{i,j} + E_{j,i})C$$

where the left (respectively, right) multiplication exchanges columns (respectively, rows) i and j of C. It follows that all the diagonal elements must be equal and so $C = cI_d$ for some $c \in \mathbf{C}$. Finally, all these matrices clearly commute with any other matrix, so we are done. ∎

Since we will be computing $Z_{\text{Com}\, X}$ and the elements of the commutant algebra involve direct sums and tensor products, we will need to know how these operations behave under multiplication.

Lemma 1.7.7 *Suppose* $A, X \in \text{Mat}_d$ *and* $B, Y \in \text{Mat}_f$. *Then*

1. $(A \oplus B)(X \oplus Y) = AX \oplus BY$,

2. $(A \otimes B)(X \otimes Y) = AX \otimes BY$.

Proof. Both assertions are easy to prove, so we will only do the second. Suppose $A = (a_{i,j})$ and $X = (x_{i,j})$. Then

$$
\begin{aligned}
(A \otimes B)(X \otimes Y) &= (a_{i,j}B)(x_{i,j}Y) & \text{(definition of } \otimes) \\
&= (\textstyle\sum_k a_{i,k}B \, x_{k,j}Y) & \text{(block multiplication)} \\
&= (\, (\textstyle\sum_k a_{i,k}x_{k,j}) \, BY) & \text{(distributivity)} \\
&= AX \otimes BY & \text{(definition of } \otimes) \; \blacksquare
\end{aligned}
$$

Now consider $C \in Z_{\text{Com}\, X}$, where X and $\text{Com}\, X$ are given by (1.12) and (1.13), respectively. So

$$CT = TC \text{ for all } T \in \text{Com}\, X \qquad (1.16)$$

where $T = \oplus_{i=1}^{k}(M_{m_i} \otimes I_{d_i})$ and $C = \oplus_{i=1}^{k}(C_{m_i} \otimes I_{d_i})$. Computing the left-hand side:

$$
\begin{aligned}
CT &= (\oplus_{i=1}^{k} C_{m_i} \otimes I_{d_i})\,(\oplus_{i=1}^{k} M_{m_i} \otimes I_{d_i}) & \text{(definition of } C \text{ and } T) \\
&= \oplus_{i=1}^{k}(C_{m_i} \otimes I_{d_i})\,(M_{m_i} \otimes I_{d_i}) & \text{(Lemma 1.7.7, item 1)} \\
&= \oplus_{i=1}^{k}(C_{m_i} M_{m_i} \otimes I_{d_i}) & \text{(Lemma 1.7.7, item 2)}.
\end{aligned}
$$

Similarly,

$$
TC = \oplus_{i=1}^{k}(M_{m_i} C_{m_i} \otimes I_{d_i})
$$

Thus equation (1.16) holds if and only if

$$
C_{m_i} M_{m_i} = M_{m_i} C_{m_i} \text{ for all } M_{m_i} \in \text{Mat}_{m_i}
$$

But this just means that C_{m_i} is in the center of Mat_{m_i}, which, by Proposition 1.7.6, is equivalent to

$$
C_{m_i} = c_i I_{m_i}
$$

for some $c_i \in \mathbf{C}$. Hence

$$
\begin{aligned}
C &= \oplus_{i=1}^{k} c_i I_{m_i} \otimes I_{d_i} \\
&= \oplus_{i=1}^{k} c_i I_{m_i d_i} \\
&= \begin{pmatrix}
c_1 I_{m_1 d_1} & 0 & \cdots & 0 \\
0 & c_2 I_{m_2 d_2} & \cdots & 0 \\
\vdots & \vdots & \ddots & \vdots \\
0 & 0 & \cdots & c_k I_{m_k d_k}
\end{pmatrix}
\end{aligned}
$$

and all members of $Z_{\text{Com} X}$ have this form. Note that $\dim Z_{\text{Com} X} = k$.

For a concrete example, let

$$
X = \begin{pmatrix}
X^{(1)} & 0 & 0 \\
0 & X^{(1)} & 0 \\
0 & 0 & X^{(2)}
\end{pmatrix} = 2X^{(1)} \oplus X^{(2)}
$$

where $\deg X^{(1)} = 3$ and $\deg X^{(2)} = 4$. Then the matrices $T \in \text{Com} X$ look like

$$
T = \left(\begin{array}{ccc|ccc|cccc}
a & 0 & 0 & b & 0 & 0 & 0 & 0 & 0 & 0 \\
0 & a & 0 & 0 & b & 0 & 0 & 0 & 0 & 0 \\
0 & 0 & a & 0 & 0 & b & 0 & 0 & 0 & 0 \\
\hline
c & 0 & 0 & d & 0 & 0 & 0 & 0 & 0 & 0 \\
0 & c & 0 & 0 & d & 0 & 0 & 0 & 0 & 0 \\
0 & 0 & c & 0 & 0 & d & 0 & 0 & 0 & 0 \\
\hline
0 & 0 & 0 & 0 & 0 & 0 & x & 0 & 0 & 0 \\
0 & 0 & 0 & 0 & 0 & 0 & 0 & x & 0 & 0 \\
0 & 0 & 0 & 0 & 0 & 0 & 0 & 0 & x & 0 \\
0 & 0 & 0 & 0 & 0 & 0 & 0 & 0 & 0 & x
\end{array}\right)
$$

where $a, b, c, d, x \in \mathbf{C}$. The dimension is evidently

$$
\dim(\text{Com} X) = m_1^2 + m_2^2 = 2^2 + 1^2 = 5
$$

The elements $C \in Z_{\text{Com } X}$ are even simpler:

$$
C = \left(\begin{array}{ccc|ccc|cccc}
a & 0 & 0 & 0 & 0 & 0 & 0 & 0 & 0 & 0 \\
0 & a & 0 & 0 & 0 & 0 & 0 & 0 & 0 & 0 \\
0 & 0 & a & 0 & 0 & 0 & 0 & 0 & 0 & 0 \\
\hline
0 & 0 & 0 & a & 0 & 0 & 0 & 0 & 0 & 0 \\
0 & 0 & 0 & 0 & a & 0 & 0 & 0 & 0 & 0 \\
0 & 0 & 0 & 0 & 0 & a & 0 & 0 & 0 & 0 \\
\hline
0 & 0 & 0 & 0 & 0 & 0 & x & 0 & 0 & 0 \\
0 & 0 & 0 & 0 & 0 & 0 & 0 & x & 0 & 0 \\
0 & 0 & 0 & 0 & 0 & 0 & 0 & 0 & x & 0 \\
0 & 0 & 0 & 0 & 0 & 0 & 0 & 0 & 0 & x
\end{array} \right)
$$

where $a, x \in \mathbf{C}$. Here the dimension is the number of different irreducible components of X, in this case 2.

We summarize these results in the following theorem.

Theorem 1.7.8 *Let X be a matrix representation of G such that*

$$X = m_1 X^{(1)} \oplus m_2 X^{(2)} \oplus \cdots \oplus m_k X^{(k)} \qquad (1.17)$$

where the $X^{(i)}$ are inequivalent, irreducible and $\deg X^{(i)} = d_i$. Then

1. $\deg X = m_1 d_1 + m_2 d_2 + \cdots + m_k d_k$,

2. $\text{Com } X = \{ \oplus_{i=1}^{k}(M_{m_i} \otimes I_{d_i}) \mid M_{m_i} \in \text{Mat}_{m_i} \text{ for all } i \}$,

3. $\dim(\text{Com } X) = m_1^2 + m_2^2 + \cdots + m_k^2$,

4. $Z_{\text{Com } X} = \{ \oplus_{i=1}^{k} c_i I_{m_i d_i} \mid c_i \in \mathbf{C} \text{ for all } i \}$, *and*

5. $\dim Z_{\text{Com } X} = k$. ∎

What happens if we try to apply Theorem 1.7.8 to a representation Y that is not decomposed into irreducibles? By the matrix version of Maschke's theorem (Corollary 1.5.4), Y is equivalent to a representation X of the form given in equation (1.17). But if $Y = RXR^{-1}$ for some fixed matrix R, then the map

$$T \rightarrow RTR^{-1}$$

is an algebra isomorphism from $\text{Com } X$ to $\text{Com } Y$. Once the commutant algebras are isomorphic, it is easy to see that their centers are too. Hence Theorem 1.7.8 continues to hold with all set equalities replaced by isomorphisms.

There is also a module version of this result. We will use the multiplicity notation for G-modules in the same way it was used for matrices.

Theorem 1.7.9 *Let V be a G-module such that*

$$V \cong m_1 V^{(1)} \oplus m_2 V^{(2)} \oplus \cdots \oplus m_k V^{(k)}$$

where the $V^{(i)}$ are pairwise inequivalent irreducibles and $\dim V^{(i)} = d_i$. Then

1. $\dim V = m_1 d_1 + m_2 d_2 + \cdots + m_k d_k$,

2. $\operatorname{End} V \cong \oplus_{i=1}^{k} \operatorname{Mat}_{m_i}$,

3. $\dim(\operatorname{End} V) = m_1^2 + m_2^2 + \cdots + m_k^2$,

4. $Z_{\operatorname{End} V}$ *is isomorphic to the algebra of diagonal matrices of degree* k, *and*

5. $\dim Z_{\operatorname{End} V} = k$. ■

The same methods can be applied to prove the following strengthening of Corollary 1.6.7 in the case where the field is $\mathbf{C}$.

Proposition 1.7.10 *Let* V *and* W *be* G-*modules with* V *irreducible. Then* $\dim \operatorname{Hom}(V, W)$ *is the multiplicity of* V *in* W. ■

1.8   Group Characters

It turns out that much of the information contained in a representation can be distilled into one simple statistic: the traces of the corresponding matrices. This is the beautiful theory of group characters that will occupy us for the rest of this chapter.

Definition 1.8.1 Let $X(g)$, $g \in G$, be a matrix representation. Then the *character of* X is
$$\chi(g) = \operatorname{tr} X(g)$$
where tr denotes the trace of a matrix. Otherwise put, χ is the map
$$G \overset{\operatorname{tr} X}{\to} \mathbf{C}$$

If V is a G-module, then its *character* is the character of a matrix representation X corresponding to V. ■

Since there are many matrix representations corresponding to a single G-module, we should check that the module character is well-defined. But if X and Y both correspond to V, then $Y = TXT^{-1}$ for some fixed T. Thus, for all $g \in G$,
$$\operatorname{tr} Y(g) = \operatorname{tr} TX(g)T^{-1} = \operatorname{tr} X(g)$$
since trace is invariant under conjugation. Hence X and Y have the same character and our definition makes sense.

Much of the terminology we have developed for representations will be applied without change to the corresponding characters. Thus if X has character χ, we will say that χ is irreducible whenever X is, etc. Now let us turn to some examples.

Example 1.8.2 Suppose G is arbitrary and X is a degree 1 representation. Then the character $\chi(g)$ is just the sole entry of $X(g)$ for each $g \in G$. Such characters are called *linear characters*. ∎

Example 1.8.3 Suppose we consider the defining representation of S_n with its character χ^{def}. If we take $n = 3$, then we can compute the character values directly by taking the traces of the matrices in Example 1.2.4. The results are

$$\chi^{\text{def}}(\ (1)(2)(3)\ ) = 3, \qquad \chi^{\text{def}}(\ (1,2)(3)\ ) = 1, \qquad \chi^{\text{def}}(\ (1,3)(2)\ ) = 1$$
$$\chi^{\text{def}}(\ (1)(2,3)\ ) = 1, \qquad \chi^{\text{def}}(\ (1,2,3)\ ) = 0, \qquad \chi^{\text{def}}(\ (1,3,2)\ ) = 0$$

It is not hard to see that, in general, if $\pi \in S_n$, then

$$
\begin{aligned}
\chi^{\text{def}}(\pi) \quad &= \quad \text{the number of ones on the diagonal of } X(\pi) \\
&= \quad \text{the number of fixed-points of } \pi \ \blacksquare
\end{aligned}
$$

Example 1.8.4 Let $G = \{g_1, g_2, \ldots, g_n\}$ and consider the regular representation with module $V = \mathbf{C}[\mathbf{G}]$ and character χ^{reg}. Now $X(\epsilon) = I_n$, so $\chi^{\text{reg}}(\epsilon) = |G|$.

To compute the character values for $g \neq \epsilon$, we will use the matrices arising from the standard basis $\mathcal{B} = \{\mathbf{g}_1, \mathbf{g}_2, \ldots, \mathbf{g}_n\}$. Now $X(g)$ is the permutation matrix for the action of g on $\mathcal{B}$, so $\chi^{\text{reg}}(g)$ is the number of fixed-points for that action. But if $g\mathbf{g}_i = \mathbf{g}_i$ for any i, then we must have $g = \epsilon$, which is not the case—i.e., there are no fixed-points if $g \neq \epsilon$. To summarize

$$\chi^{\text{reg}}(g) = \begin{cases} |G| & \text{if } g = \epsilon \\ 0 & \text{otherwise} \end{cases} \ \blacksquare$$

We now prove some elementary properties of characters.

Proposition 1.8.5 *Let X be a matrix representation of a group G of degree d with character χ.*

1. *$\chi(\epsilon) = d$.*

2. *If K is a conjugacy class of G, then*

$$g, h \in K \Rightarrow \chi(g) = \chi(h)$$

3. *If Y is a representation of G with character ψ, then*

$$X \cong Y \Rightarrow \chi(g) = \psi(g)$$

for all $g \in G$.

Proof. 1. Since $X(\epsilon) = I_d$,

$$\chi(\epsilon) = \text{tr } I_d = d$$

2. By hypothesis $g = khk^{-1}$, so

$$\chi(g) = \mathrm{tr}\, X(g) = \mathrm{tr}\, X(k)X(h)X(k)^{-1} = \mathrm{tr}\, X(h) = \chi(h)$$

3. This assertion just says that equivalent representations have the same character. We have already proved this in the remarks following the preceding definition of group characters. ∎

It is surprising that the converse of 3 is also true—i.e., if two representations have the same character, then they must be equivalent. This result (which is proved as Corollary 1.9.4, part 5) is the motivation for the paragraph with which we opened this section.

In the previous proposition, 2 says that characters are constant on conjugacy classes. Such functions have a special name.

Definition 1.8.6 A *class function* on a group G is a mapping $f : G \to \mathbf{C}$ such that $f(g) = f(h)$ whenever g and h are in the same conjugacy class. The set of all class functions on G is denoted $R(G)$.

Clearly the sums and scalar multiples of class functions are again class functions, so $R(G)$ is actually a vector space over $\mathbf{C}$. Also $R(G)$ has a natural basis consisting of those functions that have the value 1 on a given conjugacy class and 0 elsewhere. Thus

$$\dim R(G) = \text{ number of conjugacy classes of } G \qquad (1.18)$$

If K is a conjugacy class and χ is a character, we can define χ_K to be the value of the given character on the given class:

$$\chi_K = \chi(g)$$

for any $g \in K$. This brings us to the definition of the character table of a group.

Definition 1.8.7 Let G be a group. The *character table of G* is an array with rows indexed by the inequivalent irreducible characters of G and columns indexed by the conjugacy classes. The table entry in row χ and column K is χ_K:

$$
\begin{array}{c|ccc}
 & \cdots & K & \cdots \\
\hline
\vdots & & \vdots & \\
\chi & \cdots & \chi_K & \\
\vdots & & &
\end{array}
$$

By convention, the first row corresponds to the trivial character and the first column corresponds to the class of the identity, $K = \{\epsilon\}$. ∎

It is not clear that the character table is always finite: there might be an infinite number of irreducible characters of G. Fortunately, this turns out not to be the case. In fact we will prove in Section 1.10 that the number of inequivalent irreducible representations of G is equal to the number of conjugacy classes, so the character table is always square. Let's examine some examples.

Example 1.8.8 If $G = C_n$, the cyclic group with n elements, then each element of C_n is in a conjugacy class by itself (as is true for any abelian group). Since there are n conjugacy classes, there must be n inequivalent irreducible representations of C_n. But we found n degree 1 representations in Example 1.2.3, and they are pairwise inequivalent, since they all have different characters (Proposition 1.8.5, part 3). So we have found all the irreducibles for C_n.

Since the representations are one-dimensional, they are equal to their corresponding characters. Thus the table we displayed on page 5 is indeed the complete character table for C_4. ∎

Example 1.8.9 Recall that a conjugacy class in $G = \mathcal{S}_n$ consists of all permutations of a given cycle type. In particular, for $\mathcal{S}_3$ we have three conjugacy classes

$$K_1 = \{\epsilon\}, \qquad K_2 = \{(1,2),\ (1,3),\ (2,3)\}, \quad \text{and} \quad K_3 = \{(1,2,3),\ (1,3,2)\}$$

Thus there are three irreducible representations of $\mathcal{S}_3$. We've met two of them, the trivial and sign representations. So this is as much as we know of the character table for $\mathcal{S}_3$:

	K_1	K_2	K_3
$\chi^{(1)}$	1	1	1
$\chi^{(2)}$	1	−1	1
$\chi^{(3)}$	?	?	?

We will be able to fill in the last line using character inner products. ∎

1.9 Inner Products of Characters

Next, we study the powerful tool of the character inner product. Taking inner products is a simple method for determining whether a representation is irreducible. This technique will also be used to prove that equality of characters implies equivalence of representations and to show that the number of irreducibles is equal to the number of conjugacy classes. First, however, we motivate the definition.

We can think of a character χ of a group $G = \{g_1, g_2, \ldots, g_n\}$ as a row vector of complex numbers:

$$\chi = (\chi(g_1), \chi(g_2), \ldots, \chi(g_n))$$

If χ is irreducible, then this vector can be obtained from the character table by merely repeating the value for class K a total of $|K|$ times. For example, the first two characters for $\mathcal{S}_3$ in the preceding table become

$$\chi^{(1)} = (1,1,1,1,1,1) \quad \text{and} \quad \chi^{(2)} = (1,-1,-1,-1,1,1)$$

We have the usual inner product on row vectors given by

$$(c_1, c_2, \ldots, c_n) \cdot (d_1, d_2, \ldots, d_n) = c_1\overline{d_1} + c_2\overline{d_2} + \cdots + c_n\overline{d_n}$$

where the bar stands for complex conjugation. Computing with our $\mathcal{S}_3$ characters, it is easy to verify that

$$\chi^{(1)} \cdot \chi^{(1)} = \chi^{(2)} \cdot \chi^{(2)} = 6$$

and

$$\chi^{(1)} \cdot \chi^{(2)} = 0$$

More trials with other irreducible characters—e.g., those of C_4—will lead the reader to conjecture that if $\chi^{(i)}$ and $\chi^{(j)}$ are irreducible characters of G, then

$$\chi^{(i)} \cdot \chi^{(j)} = \begin{cases} |G| & \text{if } i = j \\ 0 & \text{if } i \neq j \end{cases}$$

Dividing by $|G|$ for normality gives us one definition of the character inner product.

Definition 1.9.1 Let χ and ψ be any two functions from a group G to the complex numbers $\mathbf{C}$. The *inner product* of χ and ψ is

$$\langle \chi, \psi \rangle = \frac{1}{|G|} \sum_{g \in G} \chi(g)\overline{\psi(g)} \quad \blacksquare$$

Now suppose V is a G-module with character ψ. We have seen, in the proof of Maschke's theorem, that there is an inner product on V itself that is invariant under the action of G. By picking an orthonormal basis for V, we obtain a matrix representation Y for ψ, where each $Y(g)$ is unitary; i.e.,

$$Y(g^{-1}) = Y(g)^{-1} = \overline{Y(g)}^t$$

where t denotes transpose. So

$$\overline{\psi(g)} = \operatorname{tr} \overline{Y(g)} = \operatorname{tr} Y(g^{-1})^t = \operatorname{tr} Y(g^{-1}) = \psi(g^{-1})$$

Substituting this into $\langle \cdot, \cdot \rangle$ yields another useful form of the inner product.

Proposition 1.9.2 Let χ and ψ be characters; then

$$\langle \chi, \psi \rangle = \frac{1}{|G|} \sum_{g \in G} \chi(g)\psi(g^{-1}) \quad \blacksquare \tag{1.19}$$

When the field is arbitrary, equation (1.19) is taken as the *definition* of the inner product. In fact, for any two functions χ and ψ from G to a field, we can define

$$\langle \chi, \psi \rangle' = \frac{1}{|G|} \sum_{g \in G} \chi(g)\psi(g^{-1})$$

but over the complex numbers this "inner product" is only a bilinear form. Of course, when restricted to characters we have $\langle \chi, \psi \rangle = \langle \chi, \psi \rangle'$. Also note that whenever χ and ψ are constant on conjugacy classes, we have

$$\langle \chi, \psi \rangle = \frac{1}{|G|} \sum_K |K| \chi_K \overline{\psi_K}$$

where the sum is over all conjugacy classes of G.

We can now prove that the irreducible characters are orthonormal with respect to the inner product $\langle \cdot, \cdot \rangle$.

Theorem 1.9.3 (Character Relations of the First Kind) *Let χ and ψ be irreducible characters of a group G. Then*

$$\langle \chi, \psi \rangle = \delta_{\chi,\psi}$$

Proof. Suppose χ, ψ are the characters of matrix representations A, B of degrees d, f, respectively. We will be using Schur's lemma, and so a matrix must be found to fulfill the role of T in Corollary 1.6.6. Let $X = (x_{i,j})$ be a $d \times f$ matrix of indeterminates $x_{i,j}$ and consider the matrix

$$Y = \frac{1}{|G|} \sum_{g \in G} A(g) X B(g^{-1}) \qquad (1.20)$$

We claim that $A(h)Y = YB(h)$ for all $h \in G$. Indeed,

$$
\begin{aligned}
A(h)YB(h)^{-1} &= \frac{1}{|G|} \sum_{g \in G} A(h)A(g)XB(g^{-1})B(h^{-1}) \\
&= \frac{1}{|G|} \sum_{g \in G} A(hg)XB(g^{-1}h^{-1}) \\
&= \frac{1}{|G|} \sum_{\substack{\tilde{g} \in G \\ \tilde{g}=hg}} A(\tilde{g})XB(\tilde{g}^{-1}) \\
&= Y
\end{aligned}
$$

and our assertion is proved. Thus by Corollaries 1.6.6 and 1.6.8,

$$Y = \begin{cases} 0 & \text{if } A \not\cong B \\ cI_d & \text{if } A \cong B \end{cases} \qquad (1.21)$$

Consider first the case where $\chi \neq \psi$, so that A and B must be inequivalent. Since this forces $y_{i,j} = 0$ for every element of Y, we can take the (i, j) entry of equation (1.20) to obtain

$$\frac{1}{|G|} \sum_{k,l} \sum_{g \in G} a_{i,k}(g) x_{k,l} b_{l,j}(g^{-1}) = 0$$

for all i, j. If this polynomial is to be zero, the coefficient of each $x_{k,l}$ must also be zero, so

$$\frac{1}{|G|} \sum_{g \in G} a_{i,k}(g) b_{l,j}(g^{-1}) = 0$$

for all i, j, k, l. Notice that this last equation can be more simply stated as

$$\langle a_{i,k}, b_{l,j} \rangle' = 0 \qquad \forall \ i, j, k, l \tag{1.22}$$

since our definition of inner product applies to all functions from G to $\mathbf{C}$. Now

$$\chi = \operatorname{tr} A = a_{1,1} + a_{2,2} + \cdots + a_{d,d}$$

and

$$\psi = \operatorname{tr} B = b_{1,1} + b_{2,2} + \cdots + b_{f,f}$$

so

$$\langle \chi, \psi \rangle = \langle \chi, \psi \rangle' = \sum_{i,j} \langle a_{i,i}, b_{j,j} \rangle' = 0$$

as desired.

Now suppose $\chi = \psi$. Since we are only interested in the character values, we might as well take $A = B$ also. By equation (1.21), there is a scalar $c \in \mathbf{C}$ such that $y_{i,j} = c\delta_{i,j}$. So, as in the previous paragraph, we have $\langle a_{i,k}, a_{l,j} \rangle' = 0$ as long as $i \neq j$. To take care of the case $i = j$, consider

$$\frac{1}{|G|} \sum_{g \in G} A(g) X A(g^{-1}) = cI_d$$

and take the trace on both sides

$$
\begin{aligned}
cd &= \operatorname{tr} cI_d \\
   &= \frac{1}{|G|} \sum_{g \in G} \operatorname{tr} A(g) X A(g^{-1}) \\
   &= \frac{1}{|G|} \sum_{g \in G} \operatorname{tr} X \\
   &= \operatorname{tr} X
\end{aligned}
$$

Thus $y_{i,i} = c = \frac{1}{d} \operatorname{tr} X$, which can be rewritten as

$$\frac{1}{|G|} \sum_{k,l} \sum_{g \in G} a_{i,k}(g) x_{k,l} a_{l,i}(g^{-1}) = \frac{1}{d}(x_{1,1} + x_{2,2} + \cdots + x_{d,d})$$

Equating coefficients of like monomials in this equation yields

$$\langle a_{i,k}, a_{l,i} \rangle' = \frac{1}{|G|} \sum_{g \in G} a_{i,k}(g) a_{l,i}(g^{-1}) = \frac{1}{d} \delta_{k,l} \qquad (1.23)$$

It follows that

$$
\begin{aligned}
\langle \chi, \chi \rangle &= \sum_{i,j=1}^{d} \langle a_{i,i}, a_{j,j} \rangle' \\
&= \sum_{i=1}^{d} \langle a_{i,i}, a_{i,i} \rangle' \\
&= \sum_{i=1}^{d} \frac{1}{d} \\
&= 1
\end{aligned}
$$

and the theorem is proved. ∎

Note that equations (1.22) and (1.23) give orthogonality relations for the matrix entries of the representations.

The character relations of the first kind have many interesting consequences.

Corollary 1.9.4 *Let X be a matrix representation of G with character χ. Suppose*

$$X \cong m_1 X^{(1)} \oplus m_2 X^{(2)} \oplus \cdots \oplus m_k X^{(k)}$$

where the $X^{(i)}$ are pairwise inequivalent with characters $\chi^{(i)}$.

1. $\chi = m_1 \chi^{(1)} + m_2 \chi^{(2)} + \cdots + m_k \chi^{(k)}$.

2. $\langle \chi, \chi^{(j)} \rangle = m_j$ *for all j.*

3. $\langle \chi, \chi \rangle = m_1^2 + m_2^2 + \cdots + m_k^2$.

4. X *is irreducible if and only if $\langle \chi, \chi \rangle = 1$.*

5. *Let Y be another matrix representation of G with character ψ. Then*

$$X \cong Y \text{ if and only if } \chi(g) = \psi(g)$$

for all $g \in G$.

Proof. 1. Using the fact that the trace of a direct sum is the sum of the traces:

$$\chi = \operatorname{tr} X = \operatorname{tr} \bigoplus_{i=1}^{k} m_i X^{(i)} = \sum_{i=1}^{k} m_i \chi^{(i)}$$

2. We have

$$\langle \chi, \chi^{(j)} \rangle = \langle \sum_i m_i \chi^{(i)}, \chi^{(j)} \rangle = \sum_i m_i \langle \chi^{(i)}, \chi^{(j)} \rangle = m_j$$

by the previous theorem.

3. By another application of Theorem 1.9.3:

$$\langle \chi, \chi \rangle = \langle \sum_i m_i \chi^{(i)}, \sum_j m_j \chi^{(j)} \rangle = \sum_{i,j} m_i m_j \langle \chi^{(i)}, \chi^{(j)} \rangle = \sum_i m_i^2$$

4. The assertion that X irreducible implies $\langle \chi, \chi \rangle = 1$ is just part of the orthogonality relations already proved. For the converse, suppose that

$$\langle \chi, \chi \rangle = \sum_i m_i^2 = 1$$

Then there must be exactly one index j such that $m_j = 1$ and all the rest of the m_i must be zero. But then $X = X^{(j)}$, which is irreducible by assumption.

5. The forward implication was proved as part 3 of Proposition 1.8.5. For the other direction, let $Y \cong \oplus_{i=1}^k n_i X^{(i)}$. There is no harm in assuming that the X and Y expansions both contain the same irreducibles: any irreducible found in one but not the other can be inserted with multiplicity 0. Now $\chi = \psi$, so $\langle \chi, \chi^{(i)} \rangle = \langle \psi, \chi^{(i)} \rangle$ for all i. But then, by part 2 of this corollary, $m_i = n_i$ for all i. Thus the two direct sums are equivalent—i.e., $X \cong Y$. $\blacksquare$

As an example of how these results are applied in practice, we return to the defining representation of $\mathcal{S}_n$. To simplify matters, note that both $\pi, \pi^{-1} \in \mathcal{S}_n$ have the same cycle type and are thus in the same conjugacy class. So if χ is a character of $\mathcal{S}_n$, then $\chi(\pi) = \chi(\pi^{-1})$, since characters are constant on conjugacy classes. It follows that the inner product formula for $\mathcal{S}_n$ can be rewritten as

$$\langle \chi, \psi \rangle = \frac{1}{n!} \sum_{\pi \in \mathcal{S}_n} \chi(\pi)\psi(\pi) \tag{1.24}$$

Example 1.9.5 Let $G = \mathcal{S}_3$ and consider $\chi = \chi^{\mathrm{def}}$. Let $\chi^{(1)}, \chi^{(2)}, \chi^{(3)}$ be the three irreducible characters of $\mathcal{S}_3$, where the first two are the trivial and sign characters, respectively. By Maschke's theorem, we know that

$$\chi = m_1 \chi^{(1)} + m_2 \chi^{(2)} + m_3 \chi^{(3)}$$

Furthermore, we can use equation (1.24) and part 2 of Corollary 1.9.4 to compute m_1 and m_2 (character values for $\chi = \chi^{\mathrm{def}}$ were found in Example 1.8.3):

$$m_1 = \langle \chi, \chi^{(1)} \rangle = \frac{1}{3!} \sum_{\pi \in \mathcal{S}_3} \chi(\pi)\chi^{(1)}(\pi) = \frac{1}{6}(3 \cdot 1 + 1 \cdot 1 + 1 \cdot 1 + 1 \cdot 1 + 0 \cdot 1 + 0 \cdot 1) = 1$$

$$m_2 = \langle \chi, \chi^{(2)} \rangle = \frac{1}{3!} \sum_{\pi \in \mathcal{S}_3} \chi(\pi)\chi^{(2)}(\pi) = \frac{1}{6}(3 \cdot 1 - 1 \cdot 1 - 1 \cdot 1 - 1 \cdot 1 + 0 \cdot 1 + 0 \cdot 1) = 0$$

Thus

$$\chi = \chi^{(1)} + m_3 \chi^{(3)}$$

In fact, we already knew that the defining character contained a copy of the trivial one. This was noted when we decomposed the corresponding matrices as $X = A \oplus B$, where A was the matrix of the trivial representation (see page 13). The exciting news is that the B matrices correspond to one or more copies of the mystery character $\chi^{(3)}$. These matrices turned out to be

$$B(\epsilon) = \begin{pmatrix} 1 & 0 \\ 0 & 1 \end{pmatrix}, \qquad B(\,(1,2)\,) = \begin{pmatrix} -1 & -1 \\ 0 & 1 \end{pmatrix}$$

$$B(\,(1,3)\,) = \begin{pmatrix} 1 & 0 \\ -1 & -1 \end{pmatrix}, \qquad B(\,(2,3)\,) = \begin{pmatrix} 0 & 1 \\ 1 & 0 \end{pmatrix}$$

$$B(\,(1,2,3)\,) = \begin{pmatrix} -1 & -1 \\ 1 & 0 \end{pmatrix}, \qquad B(\,(1,3,2)\,) = \begin{pmatrix} 0 & 1 \\ -1 & -1 \end{pmatrix}$$

If we let ψ be the corresponding character, then

$$\psi(\epsilon) = 2$$
$$\psi(\,(1,2)\,) = \psi(\,(1,3)\,) = \psi(\,(2,3)\,) = 0$$
$$\psi(\,(1,2,3)\,) = \psi(\,(1,3,2)\,) = -1$$

If ψ is irreducible, then $m_3 = 1$ and we have found $\chi^{(3)}$. (If not, then ψ, being of degree 2, must contain two copies of $\chi^{(3)}$.) But part 4 of Corollary 1.9.4 makes it easy to determine irreducibility; merely compute:

$$\langle \psi, \psi \rangle = \frac{1}{6}(2^2 + 0^2 + 0^2 + 0^2 + (-1)^2 + (-1)^2) = 1$$

We have found the missing irreducible. The complete character table for $\mathcal{S}_3$ is thus

	K_1	K_2	K_3
$\chi^{(1)}$	1	1	1
$\chi^{(2)}$	1	-1	1
$\chi^{(3)}$	2	0	-1

In general, the defining module for $\mathcal{S}_n$, $V = \mathbb{C}[1, 2, \ldots, n]$, always has $W = \mathbb{C}[1 + 2 + \cdots + n]$ as a submodule. If $\chi^{(1)}$ and $\chi^{\perp}$ are the characters corresponding to W and $W^{\perp}$, respectively, then $V = W \oplus W^{\perp}$. This translates to

$$\chi^{\mathrm{def}} = \chi^{(1)} + \chi^{\perp}$$

on the character level. We already know that χ^{def} counts fixed-points and that $\chi^{(1)}$ is the trivial character. Thus

$$\chi^{\perp}(\pi) = (\text{number of fixed-points of } \pi) - 1$$

is also a character of $\mathcal{S}_n$. In fact, $\chi^{\perp}$ is irreducible, although that is not obvious from the previous discussion. ∎

1.10   Decomposition of the Group Algebra

We now apply the machinery we have developed to the problem of decomposing the group algebra into irreducibles. In the process, we determine the number of inequivalent irreducible representations of any group.

Let G be a group with group algebra $\mathbf{C}[G]$ and character $\chi = \chi^{\text{reg}}$. By Maschke's theorem (Theorem 1.5.3), we can write

$$\mathbf{C}[G] = \bigoplus_i m_i V^{(i)} \tag{1.25}$$

where the $V^{(i)}$ run over all pairwise inequivalent irreducibles (and only a finite number of the m_i are nonzero).

What are the multiplicities m_i? If $V^{(i)}$ has character $\chi^{(i)}$, then, by part 2 of Corollary 1.9.4,

$$m_i = \langle \chi, \chi^{(i)} \rangle = \frac{1}{|G|} \sum_{g \in G} \chi(g) \chi^{(i)}(g^{-1})$$

But we computed the character of the regular representation in Example 1.8.4, and it vanished for $g \neq \epsilon$ with $\chi(\epsilon) = |G|$. Plugging in the preceding values:

$$m_i = \frac{1}{|G|} \chi(\epsilon) \chi^{(i)}(\epsilon) = \dim V^{(i)} \tag{1.26}$$

by Proposition 1.8.5, part 1. Hence every irreducible G-module occurs in $\mathbf{C}[G]$ with multiplicity equal to its dimension. In particular, they all appear at least once, so the list of inequivalent irreducibles must be finite (since the group algebra has finite dimension). We record these results, among others about the decomposition of $\mathbf{C}[G]$, as follows.

Proposition 1.10.1 *Let G be any finite group and suppose $\mathbf{C}[G] = \oplus_i m_i V^{(i)}$, where the $V^{(i)}$ form a complete list of pairwise inequivalent irreducible G-modules. Then*

1. $m_i = \dim V^{(i)}$,

2. $\sum_i (\dim V^{(i)})^2 = |G|$, *and*

3. *The number of $V^{(i)}$ equals the number of conjugacy classes of G.*

Proof. Part 1 is proved above, and from it, part 2 follows by taking dimensions in equation (1.25).

For part 3, recall from Theorem 1.7.9 that

$$\text{number of } V^{(i)} = \dim Z_{\text{End } \mathbf{C}[G]}$$

What do the elements of $\text{End } \mathbf{C}[G]$ look like? Given any $\mathbf{v} \in \mathbf{C}[G]$, define the map $\phi_{\mathbf{v}} : \mathbf{C}[G] \to \mathbf{C}[G]$ to be right multiplication by $\mathbf{v}$—i.e.,

$$\phi_{\mathbf{v}}(\mathbf{w}) = \mathbf{w}\mathbf{v}$$

for all $\mathbf{w} \in \mathbf{C}[\mathbf{G}]$. It is easy to verify that $\phi_{\mathbf{v}} \in \operatorname{End} \mathbf{C}[\mathbf{G}]$. In fact, these are the only elements of $\operatorname{End} \mathbf{C}[\mathbf{G}]$ because we claim that $\mathbf{C}[\mathbf{G}] \cong \operatorname{End} \mathbf{C}[\mathbf{G}]$ as vector spaces. To see this, consider $\phi : \mathbf{C}[\mathbf{G}] \to \operatorname{End} \mathbf{C}[\mathbf{G}]$ given by

$$\mathbf{v} \xrightarrow{\phi} \phi_{\mathbf{v}}$$

Proving that ϕ is linear is not hard. To show injectivity, we compute its kernel. If $\phi_{\mathbf{v}}$ is the identity map, then

$$\epsilon = \phi_{\mathbf{v}}(\epsilon) = \epsilon \mathbf{v} = \mathbf{v}$$

For surjectivity, suppose $\theta \in \operatorname{End} \mathbf{C}[\mathbf{G}]$ and consider $\theta(\epsilon)$, which is some vector $\mathbf{v}$. It follows that $\theta = \phi_{\mathbf{v}}$, because given any $g \in G$,

$$\theta(\mathbf{g}) = \theta(g\epsilon) = g\theta(\epsilon) = g\mathbf{v} = \mathbf{g}\mathbf{v} = \phi_{\mathbf{v}}(\mathbf{g})$$

and two linear transformations that agree on a basis agree everywhere. On the algebra level, our map ϕ is an anti-isomorphism, since it reverses the order of multiplication: $\phi_{\mathbf{v}}\phi_{\mathbf{w}} = \phi_{\mathbf{wv}}$ for all $\mathbf{v}, \mathbf{w} \in \mathbf{C}[\mathbf{G}]$. Thus ϕ induces an anti-isomorphism of the centers of $\mathbf{C}[\mathbf{G}]$ and $\operatorname{End} \mathbf{C}[\mathbf{G}]$ so that

$$\text{number of } V^{(i)} = \dim Z_{\mathbf{C}[\mathbf{G}]}$$

To find out what the center of the group algebra looks like, consider any $\mathbf{z} = c_1 \mathbf{g}_1 + \cdots + c_n \mathbf{g}_n \in Z_{\mathbf{C}[\mathbf{G}]}$, where the g_i are in G. Now for all $h \in G$, we have $\mathbf{z}\mathbf{h} = \mathbf{h}\mathbf{z}$ or $\mathbf{z} = \mathbf{h}\mathbf{z}\mathbf{h}^{-1}$, which can be written out as

$$c_1 \mathbf{g}_1 + \cdots + c_n \mathbf{g}_n = c_1 \mathbf{h}\mathbf{g}_1\mathbf{h}^{-1} + \cdots + c_n \mathbf{h}\mathbf{g}_n\mathbf{h}^{-1}$$

But as h takes on all possible values in G, $h g_1 h^{-1}$ runs over the conjugacy class of g_1. Since $\mathbf{z}$ remains invariant, all members of this class must have the same scalar coefficient c_1. Thus if G has k conjugacy classes $K_1, \ldots, K_k$ and we let

$$\mathbf{z}_i = \sum_{g \in K_i} \mathbf{g}, \ i = 1, \ldots, k$$

then we have shown that any $\mathbf{z} \in Z_{\mathbf{C}[\mathbf{G}]}$ can be written as

$$\mathbf{z} = \sum_{i=1}^{k} d_i \mathbf{z}_i$$

Similar considerations show that the converse holds: any linear combination of the $\mathbf{z}_i$ is in the center of $\mathbf{C}[\mathbf{G}]$. Finally we note that the set $\{\mathbf{z}_1, \ldots, \mathbf{z}_k\}$ forms a basis for $Z_{\mathbf{C}[\mathbf{G}]}$. We have already shown that they span. They must also be linearly independent, since they are sums over pairwise disjoint subsets of the basis $\{\mathbf{g} \mid g \in G\}$ of $\mathbf{C}[\mathbf{G}]$. Hence

$$\text{number of conjugacy classes } = \dim Z_{\mathbf{C}[\mathbf{G}]} = \text{number of } V^{(i)}$$

as desired. ∎

As a first application of this proposition, we derive a slightly deeper relationship between the characters and class functions

Proposition 1.10.2 *The irreducible characters of a group G form an orthonormal basis for the space of class functions R_G.*

Proof. Since the irreducible characters are orthonormal with respect to the bilinear form $\langle \cdot, \cdot \rangle$ on R_G (Theorem 1.9.3), they are linearly independent. But part 3 of Proposition 1.10.1 and equation (1.18) show that we have $\dim R_G$ such characters. Thus they are a basis. ∎

Knowing that the character table is square permits us to derive orthogonality relations for its columns as a companion to those for the rows.

Theorem 1.10.3 (Character Relations of the Second Kind) *Let K, L be conjugacy classes of G. Then*

$$\sum_\chi \chi_K \overline{\chi_L} = \frac{|G|}{|K|} \delta_{K,L}$$

where the sum is over all irreducible characters of G.

Proof. If χ and ψ are irreducible characters, then the character relations of the first kind yield

$$\langle \chi, \psi \rangle = \frac{1}{|G|} \sum_K |K| \, \chi_K \, \overline{\psi_K} = \delta_{\chi,\psi}$$

where the sum is over all conjugacy classes of G. But this says that the modified character table

$$U = \left(\sqrt{|K|/|G|} \, \chi_K \right)$$

has orthonormal rows. Hence U, being square, is a unitary matrix and has orthonormal columns. The theorem follows. ∎

As a third application of these ideas, we can now give an alternative method for finding the third line of the character table for $\mathcal{S}_3$ that does not involve actually producing the corresponding representation. Let the three irreducible characters be $X^{(1)}$, $X^{(2)}$, and $X^{(3)}$, where the first two are the trivial and sign characters, respectively. If d denotes the dimension of the corresponding module for $X^{(3)}$, then by Proposition 1.10.1, part 2,

$$1^2 + 1^2 + d^2 = |\mathcal{S}_3| = 6$$

Thus $X^{(3)}(\epsilon) = d = 2$. To find the value of $X^{(3)}$ on any other permutation, we use the orthogonality relations of the second kind. For example, to compute $x = X^{(3)}(\, (1,2) \,)$:

$$0 = \sum_{i=1}^{3} X^{(i)}(\epsilon) \overline{X^{(i)}(\, (1,2) \,)} = 1 \cdot 1 + 1(-1) + 2\overline{x}$$

so $x = 0$.

1.11 Tensor Products Again

Suppose we have representations of groups G and H and wish to construct a representation of the product group $G \times H$. It turns out that we can use the tensor product introduced in Section 1.7 for this purpose. In fact all the irreducible representations of $G \times H$ can be realized as tensor products of irreducibles for the individual groups. Proving this provides another application of the theory of characters.

Definition 1.11.1 Let G and H have matrix representations X and Y, respectively. The *tensor product representation*, $X \otimes Y$, assigns to each $(g, h) \in G \times H$ the matrix

$$(X \otimes Y)(g, h) = X(g) \otimes Y(h) \ \blacksquare$$

We must verify that this is indeed a representation. While we are at it, we might as well compute its character.

Theorem 1.11.2 *Let X and Y be matrix representations for G and H, respectively.*

1. *Then $X \otimes Y$ is a representation of $G \times H$.*

2. *If X, Y and $X \otimes Y$ have characters denoted by χ, ψ, and $\chi \otimes \psi$, respectively, then*
$$(\chi \otimes \psi)(g, h) = \chi(g)\psi(h)$$
   for all $(g, h) \in G \times H$.

Proof. 1. We verify the two conditions defining a representation. First of all

$$(X \otimes Y)(\epsilon, \epsilon) = X(\epsilon) \otimes Y(\epsilon) = I \otimes I = I$$

Secondly, if $(g, h), (g', h') \in G \times H$, then using Lemma 1.7.7, part 2,

$$
\begin{aligned}
(X \otimes Y)((g, h) \cdot (g', h')) &= (X \otimes Y)(gg', hh') \\
&= X(gg') \otimes Y(hh') \\
&= X(g)X(g') \otimes Y(h)Y(h') \\
&= (X(g) \otimes Y(h)) \cdot (X(g') \otimes Y(h')) \\
&= (X \otimes Y)(g, h) \cdot (X \otimes Y)(g', h')
\end{aligned}
$$

2. Note that for any matrices A and B,

$$\operatorname{tr} A \otimes B = \operatorname{tr}(a_{i,j}B) = \sum_i a_{i,i} \operatorname{tr} B = \operatorname{tr} A \operatorname{tr} B$$

Thus

$$(\chi \otimes \psi)(g, h) = \operatorname{tr}(X(g) \otimes Y(h)) = \operatorname{tr} X(g) \operatorname{tr} Y(h) = \chi(g)\psi(h) \ \blacksquare$$

There is a module-theoretic way to view the tensor product of representations. Let V be a G-module and W be an H-module. Then we can turn the vector space $V \otimes W$ into a $G \times H$-module by defining

$$(g, h)(\mathbf{v} \otimes \mathbf{w}) = (g\mathbf{v}) \otimes (h\mathbf{w})$$

and linearly extending the action as $\mathbf{v}$ and $\mathbf{w}$ run through bases of V and W, respectively. It is not hard to show that this definition satisfies the module axioms and is independent of the choice of bases. Furthermore, if V and W correspond to matrix representations X and Y via the basis vectors consisting of $\mathbf{v}$'s and $\mathbf{w}$'s, then $V \otimes W$ is a module for $X \otimes Y$ in the basis composed of $\mathbf{v} \otimes \mathbf{w}$'s.

Now we show how the irreducible representations of G and H completely determine those of $G \times H$.

Theorem 1.11.3 *Let G and H be groups.*

1. *If X and Y are irreducible representations of G and H, respectively, then $X \otimes Y$ is an irreducible representation of $G \times H$.*

2. *If $X^{(i)}$ and $Y^{(j)}$ are complete lists of inequivalent irreducible representations for G and H, respectively, then $X^{(i)} \otimes Y^{(j)}$ is a complete list of inequivalent irreducible $G \times H$-modules.*

Proof. 1. If ϕ is any character, then we know (Corollary 1.9.4, part 4) that the corresponding representation is irreducible if and only if $\langle \phi, \phi \rangle = 1$. Letting X and Y have characters χ and ψ, respectively, we have

$$
\begin{aligned}
\langle \chi \otimes \psi, \chi \otimes \psi \rangle &= \frac{1}{|G \times H|} \sum_{(g,h) \in G \times H} (\chi \otimes \psi)(g, h)(\chi \otimes \psi)(g^{-1}, h^{-1}) \\
&= \left[\frac{1}{|G|} \sum_{g \in G} \chi(g) \chi(g^{-1}) \right] \left[\frac{1}{|H|} \sum_{h \in H} \psi(h) \psi(h^{-1}) \right] \\
&= \langle \chi, \chi \rangle \langle \psi, \psi \rangle \\
&= 1 \cdot 1 = 1
\end{aligned}
$$

2. Let $X^{(i)}$ and $Y^{(j)}$ have characters $\chi^{(i)}$ and $\psi^{(j)}$, respectively. Then as in the proof of 1, we can show that

$$\langle \chi^{(i)} \otimes \psi^{(j)}, \chi^{(k)} \otimes \psi^{(l)} \rangle = \langle \chi^{(i)}, \chi^{(k)} \rangle \langle \psi^{(j)}, \psi^{(l)} \rangle = \delta_{i,k} \delta_{j,l}$$

Thus from Corollary 1.9.4, part 3, we see that the $\chi^{(i)} \otimes \psi^{(j)}$ are pairwise inequivalent.

To prove that the list is complete, it suffices to show that the number of such representations is the number of conjugacy classes of $G \times H$ (Proposition 1.10.1, part 3). But that quantity is the number of conjugacy classes of G times the number of conjugacy classes of H, which is the number of $X^{(i)} \otimes Y^{(j)}$. ∎

1.12   Restricted and Induced Representations

Given a group G with a subgroup H, is there a way to get representations of G from those of H or vice versa? We can answer these questions in the affirmative using the operations of restriction and induction. In fact, we have already seen an example of the latter, namely, the coset representation of Example 1.3.5.

Definition 1.12.1 Consider $H \leq G$ and a matrix representation X of G. The *restriction of X to H*, $X{\downarrow}_H^G$, is given by

$$X{\downarrow}_H^G (h) = X(h)$$

for all $h \in H$. If X has character χ, then denote the character of $X{\downarrow}_H^G$ by $\chi{\downarrow}_H^G$. ∎

It is trivial to verify that $X{\downarrow}_H^G$ is actually a representation of H. If the group G is clear from context, we will drop it in the notation and merely write $X{\downarrow}_H$. Note that even though X may be an irreducible representation of G, $X{\downarrow}_H$ need not be irreducible on H.

The process of moving from a representation of H to one of G is a bit more involved. This construction, called induction, is due to Frobenius. Suppose Y is a matrix representation of H. We might try to obtain a representation of G by merely defining Y to be zero outside of H. This won't work because singular matrices aren't allowed. However, there is a way to remedy the situation.

Definition 1.12.2 Consider $H \leq G$ and fix a transversal $t_1, \ldots, t_l$ for the left cosets of H—i.e., $G = t_1 H \uplus \cdots \uplus t_l H$, where $\uplus$ denotes disjoint union. If Y is a representation of H, then the corresponding *induced representation* $Y{\uparrow}_H^G$ assigns to each $g \in G$ the block matrix

$$Y{\uparrow}_H^G (g) = (Y(t_i^{-1} g t_j)) = \begin{pmatrix} Y(t_1^{-1} g t_1) & Y(t_1^{-1} g t_2) & \cdots & Y(t_1^{-1} g t_l) \\ Y(t_2^{-1} g t_1) & Y(t_2^{-1} g t_2) & \cdots & Y(t_2^{-1} g t_l) \\ \vdots & \vdots & \ddots & \vdots \\ Y(t_l^{-1} g t_1) & Y(t_l^{-1} g t_2) & \cdots & Y(t_l^{-1} g t_l) \end{pmatrix}$$

where $Y(g)$ is the zero matrix if $g \notin H$. ∎

We will abbreviate $Y{\uparrow}_H^G$ to $Y{\uparrow}^G$ if no confusion will result; these notations apply to the corresponding characters as well. It is not obvious that $Y{\uparrow}_H^G$ is actually a representation of G, but we will postpone that verification until after we have looked at an example.

As usual, let $G = S_3$ and consider $H = \{\epsilon, (2,3)\}$ with the transversal $G = H \uplus (1,2)H \uplus (1,3)H$ as in Example 1.6.3. Let $Y = 1$ be the trivial representation of H and consider $X = 1{\uparrow}^G$. Calculating the first row of the matrix for the transposition (1,2) yields

$$Y(\ \epsilon^{-1}(1,2)\epsilon\ ) = Y(\ (1,2)\ ) = 0 \qquad \text{(since } (1,2) \notin H)$$
$$Y(\ \epsilon^{-1}(1,2)(1,2)\ ) = Y(\ \epsilon\ ) = 1 \qquad \text{(since } \epsilon \in H)$$
$$Y(\ \epsilon^{-1}(1,2)(1,3)\ ) = Y(\ (1,3,2)\ ) = 0 \quad \text{(since } (1,3,2) \notin H)$$

Continuing in this way, we obtain

$$X(\ (1,2)\ ) = \begin{pmatrix} 0 & 1 & 0 \\ 1 & 0 & 0 \\ 0 & 0 & 1 \end{pmatrix}$$

The matrices for the coset representation are beginning to appear again. This is not an accident, as the next proposition shows.

Proposition 1.12.3 *Let $H \leq G$ have transversal $\{t_1, \ldots, t_l\}$ with cosets $\mathcal{H} = \{t_1 H, \ldots, t_l H\}$. Then the matrices of $1\uparrow_H^G$ are identical with those of G acting on the basis $\mathcal{H}$ for the coset module $\mathbf{C}[\mathcal{H}]$.*

Proof. Let the matrices for $1\uparrow^G$ and $\mathbf{C}[\mathcal{H}]$ be $X = (x_{i,j})$ and $Z = (z_{i,j})$, respectively. Both arrays contain only zeros and ones. Finally, for any $g \in G$:

$$x_{i,j}(g) = 1 \quad \Leftrightarrow \quad t_i^{-1}gt_j \in H$$
$$\Leftrightarrow \quad gt_j H = t_i H$$
$$\Leftrightarrow \quad z_{i,j}(g) = 1 \ \blacksquare$$

Thus $\mathbf{C}[\mathcal{H}]$ is a module for $1\uparrow_H^G$.

It is high time that we verified that induced representations are well-defined.

Theorem 1.12.4 *Suppose $H \leq G$ has transversal $\{t_1, \ldots, t_l\}$ and let Y be a matrix representation of H. Then $X = Y\uparrow_H^G$ is a representation of G.*

Proof. Analogous to the case where Y is the trivial representation, we prove that $X(g)$ is always a block permutation matrix; i.e., every row and column contains exactly one nonzero block $Y(t_i^{-1}gt_j)$. Consider the first column (the other cases being similar). It suffices to show that there is a unique element of H on the list $t_1^{-1}gt_1, t_2^{-1}gt_1, \ldots, t_l^{-1}gt_1$. But $gt_1 \in t_i H$ for exactly one of the t_i in our transversal, and so $t_i^{-1}gt_1 \in H$ is the element we seek.

We must first verify that $X(\epsilon)$ is the identity matrix, but that follows directly from the definition of induction.

Also, we need to show that $X(g)X(h) = X(gh)$ for all $g, h \in G$. Considering the (i, j) block on both sides, it suffices to prove

$$\sum_k Y(t_i^{-1}gt_k)Y(t_k^{-1}ht_j) = Y(t_i^{-1}ght_j)$$

For ease of notation, let $a_k = t_i^{-1}gt_k$, $b_k = t_k^{-1}ht_j$, and $c = t_i^{-1}ght_j$. Note that $a_k b_k = c$ for all k and that the sum can be rewritten

$$\sum_k Y(a_k)Y(b_k) \stackrel{?}{=} Y(c)$$

Now the proof breaks into two cases.

If $Y(c) = 0$, then $c \notin H$, and so either $a_k \notin H$ or $b_k \notin H$ for all k. Thus $Y(a_k)$ or $Y(b_k)$ is zero for each k, which forces the sum to be zero as well.

If $Y(c) \neq 0$, then $c \in H$. Let m be the unique index such that $a_m \in H$. Thus $b_m = a_m^{-1}c \in H$, and so

$$\sum_k Y(a_k)Y(b_k) = Y(a_m)Y(b_m) = Y(a_m b_m) = Y(c)$$

completing the proof. ∎

It should be noted that induction, like restriction, does not preserve irreducibility. It might also seem that an induced representation depends on the transversal and not just on the subgroup chosen. However, this is an illusion.

Proposition 1.12.5 *Consider $H \leq G$ and a matrix representation Y of H. Let $\{t_1, \ldots, t_l\}$ and $\{s_1, \ldots, s_l\}$ be two transversals for H giving rise to representation matrices X and Z, respectively, for $Y{\uparrow}^G$. Then X and Z are equivalent.*

Proof. Let χ, ψ and ϕ be the characters of X, Y and Z, respectively. Then it suffices to show that $\chi = \phi$ (Corollary 1.9.4, part 5). Now

$$\chi(g) = \sum_i \operatorname{tr} Y(t_i^{-1}gt_i) = \sum_i \psi(t_i^{-1}gt_i) \tag{1.27}$$

where $\psi(g) = 0$ if $g \notin H$. Similarly

$$\phi(g) = \sum_i \psi(s_i^{-1}gs_i)$$

Since the t_i and s_i are both transversals, we can permute subscripts if necessary to obtain $t_i H = s_i H$ for all i. Now $t_i = s_i h_i$, where $h_i \in H$ for all i, and so

$$t_i^{-1}gt_i = h_i^{-1}s_i^{-1}gs_i h_i$$

Thus $t_i^{-1}gt_i \in H$ if and only if $s_i^{-1}gs_i \in H$, and when both lie in H, they are in the same conjugacy class. It follows that $\psi(t_i^{-1}gt_i) = \psi(s_i^{-1}gs_i)$, since ψ is constant on conjugacy classes of H and zero outside. Hence the sums for χ and ϕ are the same. ∎

We next derive a useful formula for the character of an induced representation. Let H, ψ, and the t_i be as in the preceding proposition. Then $\psi(t_i^{-1}gt_i) = \psi(h^{-1}t_i^{-1}gt_i h)$ for any $h \in H$, so equation (1.27) can be rewritten as

$$\psi{\uparrow}^G (g) = \frac{1}{|H|} \sum_i \sum_{h \in H} \psi(h^{-1}t_i^{-1}gt_i h)$$

But as h runs over H and the t_i run over the transversal, the product $t_i h$ runs over all the elements of G exactly once. Thus we arrive at the identity

$$\psi{\uparrow}^G (g) = \frac{1}{|H|} \sum_{x \in G} \psi(x^{-1}gx) \tag{1.28}$$

This formula will permit us to prove the celebrated reciprocity law of Frobenius, which relates inner products of restricted and induced characters.

Theorem 1.12.6 (Frobenius Reciprocity) *Let $H \leq G$ and suppose that ψ and χ are characters of H and G, respectively. Then*

$$\langle \psi \uparrow^G, \chi \rangle = \langle \psi, \chi \downarrow_H \rangle$$

where the left inner product is calculated in G and the right one in H.

Proof. We have the following string of equalities:

$$
\begin{aligned}
\langle \psi \uparrow^G, \chi \rangle
&= \frac{1}{|G|} \sum_{g \in G} \psi \uparrow^G (g) \chi(g^{-1}) \\
&= \frac{1}{|G||H|} \sum_{x \in G} \sum_{g \in G} \psi(x^{-1}gx) \chi(g^{-1}) \quad \text{(equation (1.28))} \\
&= \frac{1}{|G||H|} \sum_{x \in G} \sum_{y \in G} \psi(y) \chi(xy^{-1}x^{-1}) \quad \text{(let } y = x^{-1}gx) \\
&= \frac{1}{|G||H|} \sum_{x \in G} \sum_{y \in G} \psi(y) \chi(y^{-1}) \quad (\chi \text{ constant on } G\text{'s classes}) \\
&= \frac{1}{|H|} \sum_{y \in G} \psi(y) \chi(y^{-1}) \quad (x \text{ constant in the sum}) \\
&= \frac{1}{|H|} \sum_{y \in H} \psi(y) \chi(y^{-1}) \quad (\psi \text{ is zero outside } H) \\
&= \langle \psi, \chi \downarrow_H \rangle \quad \blacksquare
\end{aligned}
$$

1.13 Exercises

1. An *inversion* in $\pi = x_1, x_2, \ldots, x_n \in \mathcal{S}_n$ (one-line notation) is a pair x_i, x_j such that $i < j$ and $x_i > x_j$. Let inv π be the number of inversions of π.

   a. Show that if π can be written as a product of k transpositions, then $k \equiv \text{inv } \pi \pmod 2$.

   b. Use part (a) to show that the sign of π is well-defined.

2. If group G acts on a set S and $s \in S$, then the *stabilizer of s* is $G_s = \{g \in G \mid gs = s\}$. The *orbit of s* is $\mathcal{O}_s = \{gs \mid g \in G\}$.

   a. Prove that G_s is a subgroup of G.

   b. Find a bijection between cosets of G/G_s and elements of $\mathcal{O}_s$.

   c. Show that $|\mathcal{O}_s| = |G|/|G_s|$ and use this to derive formula (1.1) for $|K_g|$.

3. Let G act on S with corresponding permutation representation $\mathbf{C}[S]$. Prove the following.

   a. The matrices for the action of G in the standard basis are permutation matrices.

   b. If the character of this representation is χ and $g \in G$, then

   $$\chi(g) = \text{the number of fixed points of } g \text{ acting on } S$$

4. Let G be an abelian group. Find all inequivalent irreducible representations of G. *Hint:* Use the fundamental theorem of abelian groups.

5. If X is a matrix representation of a group G, then its *kernel* is the set $N = \{g \in G \mid X(g) = I\}$. A representation is *faithful* if it is one-to-one.

   a. Show that N is a normal subgroup of G and find a condition on N equivalent to the representation being faithful.

   b. Suppose X has character χ and degree d. Prove that $g \in N$ if and only if $\chi(g) = d$. *Hint:* Show that $\chi(g)$ is a sum of roots of unity.

   c. Show that for the coset representation, $N = \cap_i g_i H g_i^{-1}$, where the g_i are the transversal.

   d. For each of the following representations, under what conditions are they faithful: trivial, regular, coset, sign for S_n, defining for S_n, degree 1 for C_n?

   e. Define a function Y on the group G/N by $Y(gN) = X(g)$ for $gN \in G/N$.

      i. Prove that Y is a well-defined faithful representation of G/N.
      ii. Show that Y is irreducible if and only if X is.
      iii. If X is the coset representation for a normal subgroup H of G, what is the corresponding representation Y?

6. It is possible to reverse the process of part (e) in the previous exercise. Let N be any normal subgroup of G and let Y be a representation of G/N. Define a function on G by $X(g) = Y(gN)$.

   a. Prove that X is a representation of G. We say that X has been *lifted* from the representation Y of G/N.

   b. Show that if Y is faithful, then X has kernel N.

   c. Show that X is irreducible if and only if Y is.

7. Let X be a reducible matrix representation with block form given by equation (1.4). Let V be a module for X with submodule W corresponding to A. Consider the quotient vector space $V/W = \{\mathbf{v} + W \mid \mathbf{v} \in V\}$. Show that V/W is a G-module with corresponding matrices $C(g)$. Furthermore, show that we have

$$V \cong W \oplus (V/W)$$

8. Let V be a vector space. Show that the following properties will hold in V if and only if a related property holds for a basis of V.

    a. V is a G-module.

    b. The map $\theta : V \to W$ is a G-homomorphism.

    c. The inner product $\langle \cdot, \cdot \rangle$ on V is G-invariant.

9. Why won't replacing the inner products in the proof of Maschke's theorem by bilinear forms give a demonstration valid over any field? Give a correct proof over an arbitrary field as follows. Assume that X is a reducible matrix representation of the form (1.4).

    a. Write out the equations obtained by equating blocks in $X(gh) = X(g)X(h)$. What interpretation can you give to the equations obtained from the upperleft and lowerright blocks?

    b. Use part (a) to show that

    $$TX(g)T^{-1} = \begin{pmatrix} A(g) & 0 \\ 0 & C(g) \end{pmatrix}$$

    where $T = \begin{pmatrix} I & D \\ 0 & I \end{pmatrix}$ and $D = \frac{1}{|G|} \sum_{g \in G} A(g^{-1})B(g)$.

10. Verify that the map $X : \mathbf{R}^+ \to GL_2$ given in the example at the end of Section 1.5 is a representation and that the subspace W is invariant.

11. Find $H \leq S_n$ and a set of tabloids S such that $\mathbf{C}[\mathcal{H}] \cong \mathbf{C}[S] \cong \mathbf{C}[1, 2, \ldots, n]$.

12. Let X be an irreducible matrix representation of G. Show that if $g \in Z_G$ (the center of G), then $X(g) = cI$ for some scalar c.

13. Let $\{X_1, X_2, \ldots, X_n\} \subseteq GL_d$ be a subgroup of commuting matrices. Show that these matrices are simultaneously diagonalizable using representation theory.

14. Prove the following converse of Schur's lemma. Let X be a representation of G over $\mathbf{C}$ with the property that only scalar multiples cI commute with $X(g)$ for all $g \in G$. Prove that X is irreducible.

15. Let X and Y be representations of G. The *inner tensor product*, $X \hat{\otimes} Y$, assigns to each $g \in G$ the matrix

    $$(X \hat{\otimes} Y)(g) = X(g) \otimes Y(g)$$

    a. Verify that $X \hat{\otimes} Y$ is a representation of G.

    b. Show that if X, Y, and $X \hat{\otimes} Y$ have characters denoted by χ, ψ, and $\chi \hat{\otimes} \psi$, respectively, then $(\chi \hat{\otimes} \psi)(g) = \chi(g)\psi(g)$.

   c. Find a group with irreducible representations X and Y such that $X \hat{\otimes} Y$ is *not* irreducible.

   d. However, prove that if X is of degree 1 and Y is irreducible, then so is $X \hat{\otimes} Y$.

16. Construct the character table of S_4. You may find the lifting process of Exercise 6 and the inner tensor products of Exercise 15 helpful.

17. Let D_n be the group of symmetries (rotations and reflections) of a regular n-gon. This group is called a *dihedral group*.

   a. Show that the abstract group with generators ρ, τ subject to the relations
$$\rho^n = \tau^2 = \epsilon \quad \text{and} \quad \rho\tau = \tau\rho^{-1}$$
   is isomorphic to D_n.

   b. Conclude that every element of D_n is uniquely expressible as $\tau^i \rho^j$, where $0 \le i \le 1$ and $0 \le j \le n-1$.

   c. Find the conjugacy classes of D_n.

   d. Find all the inequivalent irreducible *representations* of D_n. *Hint:* Use the fact that C_n is a normal subgroup of D_n.

18. Show that induction is transitive as follows. Suppose we have groups $G \ge H \ge K$ and a matrix representation X of K. Then
$$X\!\uparrow_K^G \cong (X\!\uparrow_K^H)\!\uparrow_H^G$$

Chapter 2

Representations of the Symmetric Group

In this chapter we construct all the irreducible representations of the symmetric group. We know that the number of such representations is equal to the number of conjugacy classes (Proposition 1.10.1), which in the case of $\mathcal{S}_n$ is the number of partitions of n. It may not be obvious how to associate an irreducible with each partition $\lambda = (\lambda_1, \lambda_2, \ldots, \lambda_l)$, but it is easy to find a corresponding subgroup $\mathcal{S}_\lambda$ that is an isomorphic copy of $\mathcal{S}_{\lambda_1} \times \mathcal{S}_{\lambda_2} \times \cdots \times \mathcal{S}_{\lambda_l}$ inside $\mathcal{S}_n$. We can now produce the right number of representations by inducing the trivial representation on each $\mathcal{S}_\lambda$ up to $\mathcal{S}_n$.

If M^λ is a module for $1 \uparrow_{\mathcal{S}_\lambda}^{\mathcal{S}_n}$, then it is too much to expect that these modules will all be irreducible. However, we will be able to find an ordering $\lambda^{(1)}, \lambda^{(2)}, \ldots$ of all partitions of n with the following nice property. The first module $M^{\lambda^{(1)}}$ will be irreducible, call it $S^{\lambda^{(1)}}$. Next, $M^{\lambda^{(2)}}$ will contain only copies of $S^{\lambda^{(1)}}$ plus a single copy of a new irreducible $S^{\lambda^{(2)}}$. In general, $M^{\lambda^{(i)}}$ will decompose into some $S^{\lambda^{(k)}}$ for $k < i$ and a unique new irreducible $S^{\lambda^{(i)}}$ called the ith Specht module. Thus the matrix giving the multiplicities for expressing $M^{\lambda^{(i)}}$ as a direct sum of the $S^{\lambda^{(j)}}$ will be lower triangular with ones down the diagonal. This immediately makes it easy to compute the irreducible characters of $\mathcal{S}_n$. We can also explicitly describe the Specht modules themselves.

Much of the material in this chapter can be found in the monograph of James [Jam 78], where the reader will find a more extensive treatment.

2.1 Young Subgroups, Tableaux, and Tabloids

Our objective in this section is to build the modules M^λ. First, however, we will introduce some notation and definitions for partitions.

If $\lambda = (\lambda_1, \lambda_2, \ldots, \lambda_l)$ is a partition of n, then we write $\lambda \vdash n$. We also use the notation $|\lambda| = \sum_i \lambda_i$, so that a partition of n satisfies $|\lambda| = n$. We can visualize λ as follows.

Definition 2.1.1 Suppose $\lambda = (\lambda_1, \lambda_2, \ldots, \lambda_l) \vdash n$. The *Ferrers diagram*, or *shape*, of λ is an array of n dots into l left-justified rows with row i containing λ_i dots for $1 \leq i \leq l$. ∎

The dot in row i and column j has coordinates (i, j), as in a matrix. Boxes (also called cells) are often used in place of dots. As an example, the partition $\lambda = (3, 3, 2, 1)$ has Ferrers diagram

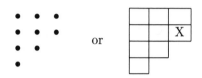

where the box in the $(2, 3)$ position has an X in it. The reader should be aware that certain authors, notably Francophones, write their Ferrers diagrams as if they were in a Cartesian coordinate system, with the first row along the x-axis, the next along the line $y = 1$, etc. With this convention, our example partition would be drawn as

but we will stick to "English" notation in this book. Also, we will use the symbol λ to stand for both the partition and its shape.

Now we wish to associate with λ a subgroup of $\mathcal{S}_n$. Recall that if A is any set, then $\mathcal{S}_A$ is the set of permutations of A.

Definition 2.1.2 Let $\lambda = (\lambda_1, \lambda_2, \ldots, \lambda_l) \vdash n$. Then the corresponding *Young subgroup* of $\mathcal{S}_n$ is

$$\mathcal{S}_\lambda = \mathcal{S}_{\{1,2,\ldots,\lambda_1\}} \times \mathcal{S}_{\{\lambda_1+1,\lambda_1+2,\ldots,\lambda_1+\lambda_2\}} \times \cdots \times \mathcal{S}_{\{n-\lambda_l+1,n-\lambda_l+2,\ldots,n\}}$$ ∎

These subgroups are named in honor of the Reverend Alfred Young, who was among the first to construct the irreducible representations of $\mathcal{S}_n$ [You 27, You 29]. For example,

$$\begin{aligned} \mathcal{S}_{(3,3,2,1)} &= \mathcal{S}_{\{1,2,3\}} \times \mathcal{S}_{\{4,5,6\}} \times \mathcal{S}_{\{7,8\}} \times \mathcal{S}_{\{9\}} \\ &\cong \mathcal{S}_3 \times \mathcal{S}_3 \times \mathcal{S}_2 \times \mathcal{S}_1 \end{aligned}$$

In general, $\mathcal{S}_{(\lambda_1,\lambda_2,\ldots,\lambda_l)}$ and $\mathcal{S}_{\lambda_1} \times \mathcal{S}_{\lambda_2} \times \cdots \times \mathcal{S}_{\lambda_l}$ are isomorphic as groups.

Now consider the representation $1{\uparrow}_{\mathcal{S}_\lambda}^{\mathcal{S}_n}$. If $\pi_1, \pi_2, \ldots, \pi_k$ is a transversal for $\mathcal{S}_\lambda$, then by Proposition 1.12.3 the vector space

$$V^\lambda = \mathbf{C}[\pi_1 \mathcal{S}_\lambda, \ \pi_2 \mathcal{S}_\lambda, \ \ldots, \ \pi_k \mathcal{S}_\lambda]$$

is a module for our induced representation. The cosets can be thought of geometrically as certain arrays in the following way.

Definition 2.1.3 Suppose $\lambda \vdash n$. A *Young tableau of shape* λ, t, is an array obtained by replacing the dots of the Ferrers diagram of λ with the numbers $1, 2, \ldots, n$ bijectively. ∎

Let $t_{i,j}$ stand for the entry of t in position (i, j). A Young tableau of shape λ is also called a λ-*tableau* and denoted by t^λ. Alternatively, we write $\operatorname{sh} t = \lambda$. Clearly there are $n!$ Young tableau for any shape $\lambda \vdash n$. To illustrate, if

$$\lambda = (2, 1) = \begin{matrix} \bullet & \bullet \\ \bullet & \end{matrix}$$

then a list of all possible tableaux of shape λ is

$$\begin{matrix} 1 & 2, \\ 3 \end{matrix} \quad \begin{matrix} 2 & 1, \\ 3 \end{matrix} \quad \begin{matrix} 1 & 3, \\ 2 \end{matrix} \quad \begin{matrix} 3 & 1, \\ 2 \end{matrix} \quad \begin{matrix} 2 & 3, \\ 1 \end{matrix} \quad \begin{matrix} 3 & 2 \\ 1 \end{matrix}$$

The first tableau above has entries $t_{1,1} = 1, t_{1,2} = 2$ and $t_{2,1} = 3$.

Young tabloids were mentioned in Example 1.6.3. We can now introduce them formally as equivalence classes of tableaux.

Definition 2.1.4 Two λ-tableaux t_1 and t_2 are *row equivalent*, $t_1 \sim t_2$, if corresponding rows of the two tableaux contain the same elements. A *tabloid of shape* λ, or λ-*tabloid*, is then

$$\{t\} = \{t_1 | t_1 \sim t\}$$

where $\operatorname{sh} t = \lambda$. ∎

As before, we will use lines between the rows of an array to indicate that it is a tabloid. So if

$$t = \begin{matrix} 1 & 2 \\ 3 \end{matrix}$$

then

$$\{t\} = \left\{ \begin{matrix} 1 & 2 \\ 3 \end{matrix}, \begin{matrix} 2 & 1 \\ 3 \end{matrix} \right\} = \begin{matrix} \underline{1 \quad 2} \\ \underline{3} \end{matrix}$$

If $\lambda = (\lambda_1, \lambda_2, \ldots, \lambda_l) \vdash n$, then the number of tableau in any given equivalence class is $\lambda_1! \lambda_2! \cdots \lambda_l! \overset{\text{def}}{=} \lambda!$. Thus the number of λ-tabloids is just $n!/\lambda!$.

Now $\pi \in \mathcal{S}_n$ acts on a tableau $t = (t_{i,j})$ of shape $\lambda \vdash n$ as follows

$$\pi t = (\pi(t_{i,j}))$$

For example,

$$(1, 2, 3) \begin{matrix} 1 & 2 \\ 3 \end{matrix} = \begin{matrix} 2 & 3 \\ 1 \end{matrix}$$

This induces an action on tabloids by letting

$$\pi \{t\} = \{\pi t\}$$

The reader should check that this is well defined—i.e., independent of the choice of t. Finally, this action gives rise, in the usual way, to an $\mathcal{S}_n$-module.

Definition 2.1.5 Suppose $\lambda \vdash n$. Let

$$M^\lambda = \mathbf{C}[\{t_1\}, \ldots, \{t_k\}]$$

where $\{t_1\}, \ldots, \{t_k\}$ is a complete list of λ-tabloids. Then M^λ is called the *permutation module corresponding to* λ. ∎

As will be seen in the next examples, we have already met three of these modules.

Example 2.1.6 If $\lambda = (n)$, then

$$M^{(n)} = \mathbf{C}[\;\boxed{\begin{array}{cccc} 1 & 2 & \cdots & n \end{array}}\;]$$

with the trivial action. ∎

Example 2.1.7 Now consider $\lambda = (1^n)$. Each equivalence class $\{t\}$ consists of a single tableau, and this tableau can be identified with a permutation in one-line notation (by taking the transpose, if you wish). Since the action of $\mathcal{S}_n$ is preserved,

$$M^{(1^n)} \cong \mathbf{C}[\mathcal{S}_n]$$

and the regular representation presents itself. ∎

Example 2.1.8 Finally, if $\lambda = (n-1, 1)$, then each λ-tabloid is uniquely determined by the element in its second row, which is a number from 1 to n. As in Example 1.6.3, this sets up a module isomorphism

$$M^{(n-1,1)} \cong \mathbf{C}[1, 2, \ldots, n]$$

so we have recovered the defining representation. ∎

Example 2.1.9 Let us compute the full set of characters of these modules when $n = 3$. In this case the only partitions are $\lambda = (3)$; $(2,1)$; and (1^3). From the previous examples, the modules M^λ correspond to the trivial, defining, and regular representations of $\mathcal{S}_3$, respectively. These character values have all been computed in Chapter 1. Denote the character of M^λ by ϕ^λ and represent the conjugacy class of $\mathcal{S}_3$ corresponding to μ by K_μ. Then we have the following table of characters.

	$K_{(1^3)}$	$K_{(2,1)}$	$K_{(3)}$
$\phi^{(3)}$	1	1	1
$\phi^{(2,1)}$	3	1	0
$\phi^{(1^3)}$	6	0	0

∎

The M^λ enjoy the following general property of modules.

Definition 2.1.10 Any G-module M is *cyclic* if there is a $\mathbf{v} \in M$ such that

$$M = \mathbf{C}[G\mathbf{v}]$$

where $G\mathbf{v} = \{g\mathbf{v} | g \in G\}$. In this case we say that M is *generated by* $\mathbf{v}$. ∎

Since any λ-tabloid can be taken to any other tabloid of the same shape by some permutation, M^λ is cyclic. We summarize in the following proposition.

Proposition 2.1.11 *If $\lambda \vdash n$, then M^λ is cyclic, generated by any given λ-tabloid. In addition, $\dim M^\lambda = n!/\lambda!$, the number of λ-tabloids.* ∎

What is the connection between our permutation modules and those obtained by inducing up from a Young subgroup? One can think of

$$\mathcal{S}_\lambda = \mathcal{S}_{\{1,2,\dots,\lambda_1\}} \times \mathcal{S}_{\{\lambda_1+1,\lambda_1+2,\dots,\lambda_1+\lambda_2\}} \times \cdots \times \mathcal{S}_{\{n-\lambda_l+1,n-\lambda_l+2,\dots,n\}}$$

as being modeled by the tabloid

$$\{t^\lambda\} = \begin{array}{|cccc|}
\hline
1 & 2 & \cdots & \lambda_1 \\
\hline
\lambda_1+1 & \lambda_1+2 & \cdots & \lambda_1+\lambda_2 \\
\hline
& \vdots & & \\
\hline
n-\lambda_l+1 & \cdots & & n \\
\hline
\end{array}$$

The fact that $\mathcal{S}_\lambda$ contains all permutations of a given interval of integers is mirrored by the fact that order of these integers is immaterial in t^λ (since they all occur in the same row). Thus the coset $\pi \mathcal{S}_\lambda$ corresponds in some way to the tabloid $\{\pi t^\lambda\}$. To be more precise, consider Theorem 2.1.12.

Theorem 2.1.12 *Consider $\lambda \vdash n$ with Young subgroup $\mathcal{S}_\lambda$ and tabloid $\{t^\lambda\}$, as before. Then $V^\lambda = \mathbf{C}[\mathcal{S}_n \mathcal{S}_\lambda]$ and $M^\lambda = \mathbf{C}[\mathcal{S}_n \{t^\lambda\}]$ are isomorphic as $\mathcal{S}_n$-modules.*

Proof. Let $\pi_1, \pi_2, \dots, \pi_k$ be a transversal for $\mathcal{S}_\lambda$. Define a map

$$\theta : V^\lambda \to M^\lambda$$

by $\theta(\pi_i \mathcal{S}_\lambda) = \{\pi_i t^\lambda\}$ for $i = 1, \dots, k$ and linear extension. It is not hard to verify that θ is the desired $\mathcal{S}_n$-isomorphism of modules. ∎

2.2 Dominance and Lexicographic Ordering

We need to find an ordering of partitions λ so that the M^λ have the nice property of the introduction to this chapter. In fact, we consider two important orderings on partitions of n, one of which is only a partial order.

Definition 2.2.1 If A is a set, then a *partial order on* A is a relation $\leq$ such that

1. $a \leq a$,

2. $a \leq b$ and $b \leq a$ implies $a = b$, and

3. $a \leq b$ and $b \leq c$ implies $a \leq c$

for all $a, b, c \in A$. In this case we say $(A, \leq)$ is a *partially ordered set*, or *poset* for short. We also write $b \geq a$ for $a \leq b$ as well as $a < b$ (or $b > a$) for $a \leq b$ and $a \neq b$.

If, in addition, for every pair of elements $a, b \in A$ we have either $a \leq b$ or $b \leq a$, then $\leq$ is a *total order* and $(A, \leq)$ is a *totally ordered set*. ∎

The three laws for a partial order are called reflexivity, antisymmetry, and transitivity, respectively. Pairs of elements $a, b \in A$ such that neither $a \leq b$ nor $b \leq a$ holds are said to be *incomparable*. One of the simplest examples of a poset is the *Boolean algebra* $B_n = (A, \subseteq)$, where A is all subsets of $\{1, 2, \ldots, n\}$ ordered by inclusion. In B_n, the subsets $\{1\}$ and $\{2\}$ are incomparable. The integers with their normal ordering are an example of a totally ordered set.

The particular partial order in which we are interested is the following.

Definition 2.2.2 Suppose $\lambda = (\lambda_1, \lambda_2, \ldots, \lambda_l)$ and $\mu = (\mu_1, \mu_2, \ldots, \mu_m)$ are partitions of n. Then λ *dominates* μ, written $\lambda \trianglerighteq \mu$, if

$$\lambda_1 + \lambda_2 + \cdots + \lambda_i \geq \mu_1 + \mu_2 + \cdots + \mu_i$$

for all $i \geq 1$. If $i > l$ (respectively, $i > m$), then we take λ_i (respectively, μ_i) to be zero. ∎

Intuitively, λ is greater than μ in the dominance order if the Ferrers diagram of λ is short and fat but the one for μ is long and skinny. For example, when $n = 6$, then $(3, 3) \trianglerighteq (2, 2, 1, 1)$, since $3 \geq 2$, $3 + 3 \geq 2 + 2$, etc. However, $(3, 3)$ and $(4, 1, 1)$ are incomparable since $3 \leq 4$ but $3 + 3 \geq 4 + 1$.

Any partially ordered set can be visualized using a Hasse diagram.

Definition 2.2.3 If $(A, \leq)$ is a poset and $b, c \in A$, then we say that b *is covered by* c, written $b \to c$, if $b < c$ and there is no $d \in A$ with $b < d < c$. The *Hasse diagram of* A consists of vertices representing the elements of A with an arrow from vertex b up to vertex c if b is covered by c. ∎

The Hasse diagram for the partitions of 6 ordered by dominance is given next.

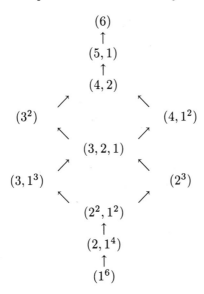

The fundamental lemma concerning the dominance order is as follows.

Lemma 2.2.4 (Dominance Lemma for Partitions) *Let t^λ and s^μ be tableaux of shape λ and μ, respectively. If, for each index i, the elements of row i of s^μ are all in different columns in t^λ, then $\lambda \trianglerighteq \mu$.*

Proof. By hypothesis, we can sort the entries in each column of t^λ so that the elements of rows $1, 2, \ldots, i$ of s^μ all occur in the first i rows of t^λ. Thus

$$
\begin{aligned}
\lambda_1 + \lambda_2 + \cdots + \lambda_i \quad &= \quad \text{number of elements in the first } i \text{ rows of } t^\lambda \\
&\geq \quad \text{number of elements of } s^\mu \text{ in the first } i \text{ rows of } t^\lambda \\
&= \quad \mu_1 + \mu_2 + \cdots + \mu_i \ \blacksquare
\end{aligned}
$$

The second ordering on partitions is the one that would be given to them in a dictionary.

Definition 2.2.5 Let $\lambda = (\lambda_1, \lambda_2, \ldots, \lambda_l)$ and $\mu = (\mu_1, \mu_2, \ldots, \mu_m)$ be partitions of n. Then $\lambda < \mu$ in *lexicographic order* if, for some index i,

$$\lambda_j = \mu_j \text{ for } j < i \text{ and } \lambda_i < \mu_i \ \blacksquare$$

This is a total ordering on partitions. For partitions of 6 we have

$$(1^6) < (2, 1^4) < (2^2, 1^2) < (2^3) < (3, 1^3) < (3, 2, 1)$$
$$< (3^2) < (4, 1^2) < (4, 2) < (5, 1) < (6)$$

The lexicographic order is a *refinement* of the dominance order in the sense of the following proposition.

Proposition 2.2.6 *If $\lambda, \mu \vdash n$ with $\lambda \trianglerighteq \mu$, then $\lambda \geq \mu$.*

Proof. If $\lambda \neq \mu$, then find the first index i where they differ. Thus $\sum_{j=1}^{i-1} \lambda_j = \sum_{j=1}^{i-1} \mu_j$ and $\sum_{j=1}^{i} \lambda_j > \sum_{j=1}^{i} \mu_j$ (since $\lambda \triangleright \mu$). So $\lambda_i > \mu_i$. ∎

It turns out that we will want to list the M^λ in *dual* lexicographic order—i.e., starting with the largest partition and working down. (By convention, the conjugacy classes are listed in the character table in the usual dictionary order so as to start with (1^n), which is the class of the identity.) Note that our first module will then be $M^{(n)}$, which is one-dimensional with the trivial action. Thus we have an irreducible module to start with, as promised.

In most of our theorems, however, we will use the dominance ordering in order to obtain the stronger conclusion that $\lambda \trianglerighteq \mu$ rather than just $\lambda > \mu$.

2.3   Specht Modules

We now construct all the irreducible modules of $\mathcal{S}_n$. These are the so-called Specht modules, S^λ.

Any tableau naturally determines certain isomorphic copies of Young subgroups in $\mathcal{S}_n$.

Definition 2.3.1 Suppose that the tableau t has rows $R_1, R_2, \ldots, R_l$ and columns $C_1, C_2, \ldots, C_k$. Then

$$R_t = \mathcal{S}_{R_1} \times \mathcal{S}_{R_2} \times \cdots \times \mathcal{S}_{R_l}$$

and

$$C_t = \mathcal{S}_{C_1} \times \mathcal{S}_{C_2} \times \cdots \times \mathcal{S}_{C_k}$$

are the *row-stabilizer* and *column-stabilizer* of t, respectively. ∎

If we take

$$t = \begin{array}{ccc} 4 & 1 & 2 \\ 3 & 5 & \end{array} \tag{2.1}$$

then

$$R_t = \mathcal{S}_{\{1,2,4\}} \times \mathcal{S}_{\{3,5\}}$$

and

$$C_t = \mathcal{S}_{\{3,4\}} \times \mathcal{S}_{\{1,5\}} \times \mathcal{S}_{\{2\}}$$

Note that our equivalence classes can be expressed as $\{t\} = R_t t$. In addition, these groups are associated with certain elements of $\mathbf{C}[\mathcal{S}_n]$. In general, given a subset $H \subseteq \mathcal{S}_n$, we can form the group algebra sums

$$H^+ = \sum_{\pi \in H} \pi$$

and

$$H^- = \sum_{\pi \in H} \mathrm{sgn}(\pi)\pi$$

(Here, the group algebra will be acting on the permutation modules M^λ. Thus the elements of $\mathbf{C}[S_n]$ are not playing the roles of vectors and are not set in boldface.) For a tableau t, the element R_t^+ is already implicit in the corresponding tabloid by the remark at the end of the previous paragraph. However, we will also need to make use of

$$\kappa_t \stackrel{\text{def}}{=} C_t^- = \sum_{\pi \in C_t} \text{sgn}(\pi)\pi$$

Note that if t has columns $C_1, C_2, \ldots, C_k$, then κ_t factors as

$$\kappa_t = \kappa_{C_1} \kappa_{C_2} \cdots \kappa_{C_k}$$

Finally, we can pass from t to an element of the module M^λ by the following definition.

Definition 2.3.2 If t is a tableau, then the associated *polytabloid* is

$$e_t = \kappa_t \{t\} \blacksquare$$

To illustrate these concepts using the tableau t in (2.1), we compute

$$\kappa_t = (\epsilon - (3,4))(\epsilon - (1,5))$$

and so

$$
e_t = \begin{array}{|c|c|c|}\hline 4 & 1 & 2 \\\hline 3 & 5 \\\cline{1-2}\end{array} - \begin{array}{|c|c|c|}\hline 3 & 1 & 2 \\\hline 4 & 5 \\\cline{1-2}\end{array} - \begin{array}{|c|c|c|}\hline 4 & 5 & 2 \\\hline 3 & 1 \\\cline{1-2}\end{array} + \begin{array}{|c|c|c|}\hline 3 & 5 & 2 \\\hline 4 & 1 \\\cline{1-2}\end{array}
$$

The next lemma describes what happens to the various objects defined above when passing from t to πt.

Lemma 2.3.3 *Let t be a tableau and π be a permutation. Then*

1. $R_{\pi t} = \pi R_t \pi^{-1}$,

2. $C_{\pi t} = \pi C_t \pi^{-1}$,

3. $\kappa_{\pi t} = \pi \kappa_t \pi^{-1}$,

4. $e_{\pi t} = \pi e_t$.

Proof. 1. We have the following list of equivalent statements.

$$
\begin{aligned}
\sigma \in R_{\pi t} &\longleftrightarrow \sigma\{\pi t\} = \{\pi t\} \\
&\longleftrightarrow \pi^{-1}\sigma\pi\{t\} = \{t\} \\
&\longleftrightarrow \pi^{-1}\sigma\pi \in R_t \\
&\longleftrightarrow \sigma \in \pi R_t \pi^{-1}
\end{aligned}
$$

The proofs of parts 2 and 3 are similar to that of part 1.

4. We have

$$e_{\pi t} = \kappa_{\pi t}\{\pi t\} = \pi\kappa_t\pi^{-1}\{\pi t\} = \pi\kappa_t\{t\} = \pi e_t \ \blacksquare$$

One can think of this lemma, for example part 1, as follows. If t has entries $t_{i,j}$, then πt has entries $\pi t_{i,j}$. Thus an element of the row-stabilizer of πt may be constructed by first applying π^{-1} to obtain the tableau t, then permuting the elements within each row of t, and finally reapplying π to restore the correct labels. Alternatively, we've just shown that the following diagram commutes.

$$
\begin{array}{ccc}
 & R_{\pi t} & \\
\pi t & \longrightarrow & \pi t_1 \\
\pi^{-1} \downarrow & & \uparrow \ \pi \\
t & \longrightarrow & t_1 \\
 & R_t &
\end{array}
$$

Finally, we are in a position to define the Specht modules.

Definition 2.3.4 For any partition λ, the corresponding *Specht module, S^λ,* is the submodule of M^λ spanned by the polytabloids e_t, where t is of shape λ. $\blacksquare$

Because of part 4 of Lemma 2.3.3, we have the following.

Proposition 2.3.5 *The S^λ are cyclic modules generated by any given polytabloid.* $\blacksquare$

Let us look at some examples.

Example 2.3.6 Suppose $\lambda = (n)$. Then $e_{1\ 2\ \cdots\ n} = \overline{1\ 2\ \cdots\ n}$ is the only polytabloid and $S^{(n)}$ carries the trivial representation. This is, of course, the only possibility, since $S^{(n)}$ is a submodule of $M^{(n)}$ where $\mathcal{S}_n$ acts trivially (Example 2.1.6). $\blacksquare$

Example 2.3.7 Let $\lambda = (1^n)$ and fix

$$
t = \begin{array}{c} 1 \\ 2 \\ \vdots \\ n \end{array} \tag{2.2}
$$

Thus

$$\kappa_t = \sum_{\sigma \in \mathcal{S}_n} (\operatorname{sgn}\sigma)\sigma$$

and e_t is the signed sum of all $n!$ permutations regarded as tabloids. Now for any permutation π, Lemma 2.3.3 yields

$$e_{\pi t} = \pi e_t = \sum_{\sigma \in \mathcal{S}_n} (\operatorname{sgn}\sigma)\pi\sigma\{t\}$$

Replacing $\pi\sigma$ by τ,

$$e_{\pi t} = \sum_{\tau \in S_n} (\operatorname{sgn} \pi^{-1}\tau)\tau\{t\} = (\operatorname{sgn} \pi^{-1}) \sum_{\tau \in S_n} (\operatorname{sgn} \tau)\tau\{t\} = (\operatorname{sgn} \pi)e_t$$

because $\operatorname{sgn} \pi^{-1} = \operatorname{sgn} \pi$. Thus every polytabloid is a scalar multiple of e_t, where t is given by equation (2.2). So

$$S^{(1^n)} = \mathbf{C}[e_t]$$

with the action $\pi e_t = (\operatorname{sgn} \pi)e_t$. This is the sign representation. ■

Example 2.3.8 If $\lambda = (n-1, 1)$, then we can use the module isomorphism of Example 2.1.8 to abuse notation and write $(n-1, 1)$-tabloids as

$$\{t\} = \frac{\begin{array}{ccc} \mathbf{i} & \cdots & \mathbf{j} \\ \hline \mathbf{k} \end{array}}{} \stackrel{\text{def}}{=} \mathbf{k}$$

This tabloid has $e_t = \mathbf{k} - \mathbf{i}$, and the span of all such vectors is easily seen to be

$$S^{(n-1,1)} = \{c_1\mathbf{1} + c_2\mathbf{2} + \cdots + c_n\mathbf{n} \mid c_1 + c_2 + \cdots + c_n = 0\}$$

So $\dim S^{(n-1,1)} = n - 1$ and we can choose a basis for this module; e.g.,

$$\mathcal{B} = \{\mathbf{2} - \mathbf{1}, \mathbf{3} - \mathbf{1}, \dots, \mathbf{n} - \mathbf{1}\}$$

Computing the action of $\pi \in S_n$ on $\mathcal{B}$, we see that the corresponding character is one less than the number of fixed-points of π. Thus $S^{(n-1,1)}$ is the module found at the end of Example 1.9.5. ■

The reader should verify that when $n = 3$, the preceding three examples do give all the irreducible representations of S_3 (which were found in Chapter 1 by other means). So at least in that case, we have fulfilled our aim.

2.4 The Submodule Theorem

It is time to show that the S^λ constitute a full set of irreducible S_n-modules. The crucial result will be the submodule theorem of James [Jam 76]. All the results and proofs of this section, up to and including this theorem, are true over an arbitrary field. The only change needed is to substitute a bilinear form for the inner product of equation (2.3). The fact that this replaces linearity by conjugate linearity in the second variable is not a problem, since we never need to carry a nonreal scalar from one side to the other.

Recall that $H^- = \sum_{\pi \in H}(\operatorname{sgn} \pi)\pi$ for any subset $H \subseteq S_n$. If $H = \{\pi\}$, then we write π^- for H^-. We will also need the unique inner product on M^λ for which

$$\langle \{t\}, \{s\} \rangle = \delta_{\{t\},\{s\}} \tag{2.3}$$

Lemma 2.4.1 (Sign Lemma) *Let $H \leq S_n$ be a subgroup.*

1. *If $\pi \in H$, then*
$$\pi H^- = H^- \pi = (\operatorname{sgn} \pi) H^-$$
Otherwise put: $\pi^- H^- = H^-$.

2. *For any $\mathbf{u}, \mathbf{v} \in M^\lambda$,*
$$\langle H^- \mathbf{u}, \mathbf{v} \rangle = \langle \mathbf{u}, H^- \mathbf{v} \rangle$$

3. *If the transposition $(b, c) \in H$, then we can factor*
$$H^- = k(\epsilon - (b, c))$$
where $k \in \mathbf{C}[S_n]$.

4. *If t is a tableau with b, c in the same row of t and $(b, c) \in H$, then*
$$H^- \{t\} = 0$$

Proof. 1. This is just like the proof that $\pi e_t = (\operatorname{sgn} \pi) e_t$ in Example 2.3.7.

2. Using the fact that our form is S_n-invariant
$$\langle H^- \mathbf{u}, \mathbf{v} \rangle = \sum_{\pi \in H} \langle (\operatorname{sgn} \pi) \pi \mathbf{u}, \mathbf{v} \rangle = \sum_{\pi \in H} \langle \mathbf{u}, (\operatorname{sgn} \pi) \pi^{-1} \mathbf{v} \rangle$$

Replacing π by π^{-1} and noting that this does not affect the sign, we see that this last sum equals $\langle \mathbf{u}, H^- \mathbf{v} \rangle$.

3. Consider the subgroup $K = \{\epsilon, (b, c)\}$ of H. Then we can find a transversal and write $H = \biguplus_i k_i K$. But then $H^- = (\sum_i k_i^-)(\epsilon - (b, c))$, as desired.

4. By hypothesis, $(b, c)\{t\} = \{t\}$. Thus
$$H^- \{t\} = k(\epsilon - (b, c))\{t\} = k(\{t\} - \{t\}) = 0 \blacksquare$$

The sign lemma has a couple of useful corollaries.

Corollary 2.4.2 *Let $t = t^\lambda$ be a λ-tableau and $s = s^\mu$ be a μ-tableau, where $\lambda, \mu \vdash n$. If $\kappa_t \{s\} \neq 0$, then $\lambda \trianglerighteq \mu$. And if $\lambda = \mu$, then $\kappa_t \{s\} = \pm e_t$.*

Proof. Suppose b and c are two elements in the same row of s^μ. Then they cannot be in the same column of t^λ, for if so, then $\kappa_t = k(\epsilon - (b, c))$ and $\kappa_t \{s\} = 0$ by parts 3 and 4 in the preceding lemma. Thus the dominance lemma (Lemma 2.2.4) yields $\lambda \trianglerighteq \mu$.

If $\lambda = \mu$, then we must have $\{s\} = \pi\{t\}$ for some $\pi \in C_t$ by the same argument that established the dominance lemma. Using part 1 yields
$$\kappa_t \{s\} = \kappa_t \pi \{t\} = (\operatorname{sgn} \pi) \kappa_t \{t\} = \pm e_t \blacksquare$$

Corollary 2.4.3 *If* $\mathbf{u} \in M^\mu$ *and* $\operatorname{sh} t = \mu$, *then* $\kappa_t \mathbf{u}$ *is a multiple of* e_t.

Proof. We can write $\mathbf{u} = \sum_i c_i \{s_i\}$, where the s_i are μ-tableaux. By the previous corollary, $\kappa_t \mathbf{u} = \sum_i \pm c_i e_t$. ∎

We are now in a position to prove the submodule theorem.

Theorem 2.4.4 (Submodule Theorem [Jam 76]) *Let U be a submodule of M^μ. Then*
$$U \supseteq S^\mu \quad or \quad U \subseteq S^{\mu\perp}$$
In particular, when the field is **C***, the S^μ are irreducible.*

Proof. Consider $\mathbf{u} \in U$ and a μ-tableau t. By the preceding corollary, we know that $\kappa_t \mathbf{u} = f e_t$ for some field element f. There are two cases, depending on which multiples can arise.

Suppose that there exits a $\mathbf{u}$ and a t with $f \neq 0$. Then since $\mathbf{u}$ is in the submodule U, we have $f e_t = \kappa_t \mathbf{u} \in U$. Thus $e_t \in U$ (since f is nonzero) and $S^\mu \subseteq U$ (since S^μ is cyclic).

On the other hand, suppose we always have $\kappa_t \mathbf{u} = \mathbf{0}$. We claim that this forces $U \subseteq S^{\mu\perp}$. Consider any $\mathbf{u} \in U$. Given an arbitrary μ-tableau t, we can apply part 2 of the sign lemma to obtain

$$\langle \mathbf{u}, e_t \rangle = \langle \mathbf{u}, \kappa_t \{t\} \rangle = \langle \kappa_t \mathbf{u}, \{t\} \rangle = \langle \mathbf{0}, \{t\} \rangle = 0$$

Since the e_t span S^μ, we have $\mathbf{u} \in S^{\mu\perp}$, as claimed. ∎

It is only now that we will need our field to be the complexes (or any field of characteristic 0).

Proposition 2.4.5 *Suppose the field of scalars is* **C** *and* $\theta \in \operatorname{Hom}(S^\lambda, M^\mu)$ *is nonzero. Thus* $\lambda \trianglerighteq \mu$ *and if* $\lambda = \mu$, *then* θ *is multiplication by a scalar.*

Proof. Since $\theta \neq 0$, there is some basis vector e_t such that $\theta(e_t) \neq \mathbf{0}$. Because $\langle \cdot, \cdot \rangle$ is an inner product with complex scalars, $M^\lambda = S^\lambda \oplus S^{\lambda\perp}$. Thus we can extend θ to an element of $\operatorname{Hom}(M^\lambda, M^\mu)$ by setting $\theta(S^{\lambda\perp}) = \mathbf{0}$. So

$$\mathbf{0} \neq \theta(e_t) = \theta(\kappa_t \{t\}) = \kappa_t \theta(\{t\}) = \kappa_t \left(\sum_i c_i \{s_i\} \right)$$

where the s_i are μ-tableaux. By Corollary 2.4.2 we have $\lambda \trianglerighteq \mu$.

In the case $\lambda = \mu$, Corollary 2.4.3 yields $\theta(e_t) = c e_t$ for some constant c. So for any permutation π,

$$\theta(e_{\pi t}) = \theta(\pi e_t) = \pi \theta(e_t) = \pi(c e_t) = c e_{\pi t}$$

Thus θ is multiplication by c. ∎

We can finally verify all our claims about the Specht modules.

Theorem 2.4.6 *The S^λ for $\lambda \vdash n$ form a complete list of irreducible $\mathcal{S}_n$-modules over the complex field.*

Proof. The S^λ are irreducible by the submodule theorem and the fact that $S^\lambda \cap S^{\lambda\perp} = 0$ for the field $\mathbf{C}$.

Since we have the right number of modules for a full set, it suffices to show that they are pairwise inequivalent. But if $S^\lambda \cong S^\mu$, then there is a nonzero homomorphism $\theta \in \text{Hom}(S^\lambda, M^\mu)$, since $S^\mu \subseteq M^\mu$. Thus $\lambda \trianglerighteq \mu$ (Proposition 2.4.5). Similarly, $\mu \trianglerighteq \lambda$, so $\lambda = \mu$. ∎

Although the Specht modules are not necessarily irreducible over a field of characteristic p, p prime, the submodule theorem says that the quotient $S^\lambda/(S^\lambda \cap S^{\lambda\perp})$ is. These are the objects that play the role of S^λ in the theory of p-modular representations of $\mathcal{S}_n$. See James [Jam 78] for further details.

Corollary 2.4.7 *The permutation modules decompose as*

$$M^\mu = \bigoplus_{\lambda \trianglerighteq \mu} m_{\lambda\mu} S^\lambda$$

with the diagonal multiplicity $m_{\mu\mu} = 1$.

Proof. This result follows from Proposition 2.4.5. If S^λ appears in M^μ with nonzero coefficient, then $\lambda \trianglerighteq \mu$. If $\lambda = \mu$, then we can also apply Proposition 1.7.10 to obtain

$$m_{\mu\mu} = \dim \text{Hom}(S^\mu, M^\mu) = 1 \; ∎$$

The coefficients $m_{\lambda\mu}$ have a nice combinatorial interpretation, as we will see in Section 2.11.

2.5 Standard Tableaux and a Basis for S^λ

In general, the polytabloids that generate S^λ are not independent. It would be nice to have a subset forming a basis—e.g., for computing the matrices and characters of the representation. There is a very natural set of tableau that can be used to index a basis.

Definition 2.5.1 A tableau t is *standard* if the rows and columns of t are increasing sequences. In this case we also say that the corresponding tabloid and polytabloid are standard. ∎

For example,

$$t = \begin{array}{ccc} 1 & 2 & 3 \\ 4 & 6 & \\ 5 & & \end{array}$$

is standard, but

$$t = \begin{array}{ccc} 1 & 2 & 3 \\ 5 & 4 & \\ 6 & & \end{array}$$

is not.

The next theorem is true over an arbitrary field.

Theorem 2.5.2 *The set*

$$\{e_t \mid t \text{ is a standard } \lambda\text{-tableau }\}$$

is a basis for S^λ.

We will spend the next two sections proving this theorem. First we will establish that the e_t are independent. As before, we will need a partial order, this time on tabloids.

It is convenient at this point to consider ordered partitions.

Definition 2.5.3 A *composition of n* is an ordered sequence of nonnegative integers

$$\lambda = (\lambda_1, \lambda_2, \ldots, \lambda_l)$$

such that $\sum_i \lambda_i = n$. The integers λ_i are called the *parts* of the composition.
∎

Note that there is no weakly decreasing condition on the parts in a composition. Thus $(1,3,2)$ and $(3,2,1)$ are both compositions of 6, but only the latter is a partition. The definitions of a Ferrers diagram and tableau are extended to compositions in the obvious way. (However, there are no standard λ-tableaux if λ is not a partition, since places to the right of or below the shape of λ are considered to be filled with an infinity symbol. Thus columns do not increase for a general composition.) The dominance order on compositions is defined exactly as in Definition 2.2.2, only now $\lambda_1, \ldots, \lambda_i$ and $\mu_1, \ldots, \mu_i$ are just the first i parts of their respective compositions, not necessarily the i largest.

Now suppose that $\{t\}$ is a tabloid with $\operatorname{sh} t = \lambda \vdash n$. For each index i, $1 \leq i \leq n$, let

$$\{t^i\} = \text{ the tabloid formed by all elements} \leq i \text{ in } \{t\}$$

and

$$\lambda^i = \text{ the composition which is the shape of } \{t^i\}$$

As an example, consider

$$\{t\} = \begin{array}{|c|c|} \hline 2 & 4 \\ \hline 1 & 3 \\ \hline \end{array}$$

Then

$$\{t^1\} = \begin{array}{|c|} \hline \emptyset \\ \hline 1 \\ \hline \end{array}, \qquad \{t^2\} = \begin{array}{|c|} \hline 2 \\ \hline 1 \\ \hline \end{array}, \qquad \{t^3\} = \begin{array}{|c|c|} \hline 2 \\ \hline 1 & 3 \\ \hline \end{array}, \qquad \{t^4\} = \begin{array}{|c|c|} \hline 2 & 4 \\ \hline 1 & 3 \\ \hline \end{array}$$

$$\lambda^1 = (0,1), \qquad \lambda^2 = (1,1), \qquad \lambda^3 = (1,2), \qquad \lambda^4 = (2,2)$$

The dominance order on tabloids is determined by the dominance ordering on the corresponding compositions.

Definition 2.5.4 Let $\{s\}$ and $\{t\}$ be two tabloids with composition sequences λ^i and μ^i, respectively. Then $\{s\}$ *dominates* $\{t\}$, written $\{s\} \trianglerighteq \{t\}$, if $\lambda^i \trianglerighteq \mu^i$ for all i. ■

The Hasse diagram for this ordering of the $(2,2)$-tabloids is as follows.

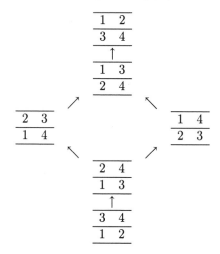

Just as for partitions, there is a dominance lemma for tabloids.

Lemma 2.5.5 (Dominance Lemma for Tabloids) *If $k < l$ and k appears in a lower row than l in $\{t\}$, then*

$$\{t\} \triangleleft (k,l)\{t\}$$

Proof. Suppose that $\{t\}$ and $(k,l)\{t\}$ have composition sequences λ^i and μ^i, respectively. Then for $i < k$ or $i \geq l$ we have $\lambda^i = \mu^i$.

Now consider the case where $k \leq i < l$. If r and q are the rows of $\{t\}$ in which k and l appear, respectively, then

$$\lambda^i = \mu^i \text{ with the } q^{\text{th}} \text{ part decreased by 1}$$
$$\text{and the } r^{\text{th}} \text{ part increased by 1}$$

Since $q < r$ by assumption, $\lambda^i \triangleright \mu^i$. ■

If $\mathbf{v} = \sum_i c_i\{t_i\} \in M^\mu$, then we say that $\{t_i\}$ *appears in* $\mathbf{v}$ if $c_i \neq 0$. The dominance lemma puts a restriction on which tableaux can appear in a standard polytabloid.

Corollary 2.5.6 *If t is standard and $\{s\}$ appears in e_t, then $\{t\} \trianglerighteq \{s\}$.*

Proof. Let $s = \pi t$, where $\pi \in C_t$, so that $\{s\}$ appears in e_t. We induct on the number of *column inversions* in s—i.e., the number of pairs $k < l$ in the same column of s such that k is in a lower row than l. Given any such pair,

$$\{s\} \triangleleft (k,l)\{s\}$$

by Lemma 2.5.5. But $(k,l)\{s\}$ has fewer inversions than $\{s\}$, so, by induction, $(k,l)\{s\} \trianglelefteq \{t\}$ and the result follows. ∎

The previous corollary says that $\{t\}$ is the maximum tabloid in e_t, by which we mean the following.

Definition 2.5.7 Let $(A, \leq)$ be a poset. Then an element $b \in A$ is the *maximum* if $b \geq c$ for all $c \in A$. An element b is a *maximal* element if there is no $c \in A$ with $c > b$. ∎

Thus a maximum element is maximal, but the converse is not necessarily true. It is important to keep this distinction in mind in the next result.

Lemma 2.5.8 *Let* $\mathbf{v}_1, \mathbf{v}_2, \ldots, \mathbf{v}_m$ *be elements of* M^μ. *Suppose, for each* $\mathbf{v}_i$, *we can choose a tabloid* $\{t_i\}$ *appearing in* $\mathbf{v}_i$ *such that*

1. $\{t_i\}$ *is maximum in* $\mathbf{v}_i$, *and*

2. *the* $\{t_i\}$ *are all distinct.*

Then $\mathbf{v}_1, \mathbf{v}_2, \ldots, \mathbf{v}_m$ *are independent.*

Proof. Choose the labels such that $\{t_1\}$ is maximal among the $\{t_i\}$. Thus conditions 1 and 2 ensure that $\{t_1\}$ appears only in $\mathbf{v}_1$. (If $\{t_1\}$ occurs in $\mathbf{v}_i$, $i > 1$, then $\{t_1\} \triangleleft \{t_i\}$, contradicting the choice of $\{t_1\}$.) It follows that in any linear combination

$$c_1\mathbf{v}_1 + c_2\mathbf{v}_2 + \cdots + c_m\mathbf{v}_m = 0$$

we must have $c_1 = 0$ because there is no other way to cancel $\{t_1\}$. By induction on m, the rest of the coefficients must also be zero. ∎

The reader should note two things about this lemma. First of all, it is not sufficient only to have the $\{t_i\}$ maximal in $\mathbf{v}_i$; it is easy to construct counterexamples. Also, we have used no special properties of $\trianglerighteq$ in the proof, so the result remains true for any partial order on tabloids.

We now have all the ingredients to prove independence of the standard basis.

Proposition 2.5.9 *The set*

$$\{e_t \mid t \text{ is a standard } \lambda\text{-tableau}\}$$

is independent.

Proof. By Corollary 2.5.6, $\{t\}$ is maximum in e_t, and by hypothesis they are all distinct. Thus Lemma 2.5.8 applies. ∎

2.6 Garnir Elements

To show that the standard polytabloids of shape μ span S^μ, we use a procedure known as a *straightening algorithm*. The basic idea is this. Pick an arbitrary tableau t. We must show that e_t is a linear combination of standard polytabloids. We may as well assume that the columns of t are increasing, since, if not, there is $\sigma \in C_t$ such that $s = \sigma t$ has increasing columns. So

$$e_s = \sigma e_t = (\text{sgn}\,\sigma)e_t$$

by Lemmas 2.3.3 (part 4) and 2.4.1 (part 1). Thus e_t is is a linear combination of polytabloids whenever e_s is.

Now suppose we can find permutations π such that

1. in each tableaux πt, a certain *row descent* of t (pair of adjacent, out-of-order elements in a row) has been eliminated, and

2. the group algebra element $g = \epsilon + \sum_\pi \pi$ satisfies $g e_t = 0$.

Then

$$e_t = -\sum_\pi e_{\pi t}$$

So we've expressed e_t in terms of polytabloids that are closer to being standard, and induction applies to obtain e_t as a linear combination of polytabloids.

The elements of the group algebra that accomplish this task are the Garnir elements.

Definition 2.6.1 Let A and B be two disjoint sets of positive integers and choose permutations π such that

$$\mathcal{S}_{A \cup B} = \biguplus_\pi \pi(\mathcal{S}_A \times \mathcal{S}_B)$$

Then a corresponding *Garnir element* is

$$g_{A,B} = \sum_\pi (\text{sgn}\,\pi)\pi \quad \blacksquare$$

Although $g_{A,B}$ depends on the transversal and not just on A and B, we will standardize the choice of the π's in a moment. Perhaps the simplest way to obtain a transversal is as follows. The group $\mathcal{S}_{A \cup B}$ acts on all ordered pairs (A', B') such that $|A'| = |A|$, $|B'| = |B|$, and $A' \uplus B' = A \uplus B$ in the obvious manner. If, for each possible (A', B'), we take $\pi \in \mathcal{S}_{A \cup B}$ such that

$$\pi(A, B) = (A', B')$$

then the collection of such permutations forms a transversal. For example, suppose $A = \{5,6\}$ and $B = \{2,4\}$. Then the corresponding pairs of sets (set brackets and commas having been eliminated for readability) and possible permutations are

$$(A', B') : (56, 24) , (46, 25) , (26, 45) , (45, 26) , (25, 46) , \quad (24, 56)$$
$$g_{A,B} = \quad \epsilon \quad - (4, 5) - (2, 5) - (4, 6) - (2, 6) + (2, 5)(4, 6)$$

It should be emphasized that for any given pair (A', B'), there are many different choices for the permutation π sending (A, B) to that pair.

The Garnir element associated with a tableau t is used to eliminate a descent $t_{i,j} > t_{i,j+1}$.

Definition 2.6.2 Let t be a tableau and let A and B be subsets of the jth and $(j+1)^{\text{st}}$ columns of t, respectively. The *Garnir element associated with t* (and A, B) is $g_{A,B} = \sum_{\pi} (\text{sgn } \pi)\pi$, where the π have been chosen so that the elements of $A \cup B$ are increasing down the columns of πt. ∎

In practice, we always take A (respectively, B) to be all elements below $t_{i,j}$ (respectively, above $t_{i,j+1}$), as in the diagram

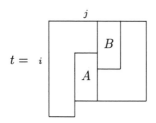

If we use the tableau

$$t = \begin{array}{ccc} 1 & 2 & 3 \\ 5 & 4 & \\ 6 & & \end{array}$$

with the descent $5 > 4$, then the sets A and B are the same as in the previous example. Each (A', B') has a corresponding t' that determines a permutation in $g_{A,B}$:

$$t' : \begin{array}{ccc} 1 & 2 & 3 \\ 5 & 4 & \\ 6 & & \end{array} , \quad \begin{array}{ccc} 1 & 2 & 3 \\ 4 & 5 & \\ 6 & & \end{array} , \quad \begin{array}{ccc} 1 & 4 & 3 \\ 2 & 5 & \\ 6 & & \end{array} , \quad \begin{array}{ccc} 1 & 2 & 3 \\ 4 & 6 & \\ 5 & & \end{array} , \quad \begin{array}{ccc} 1 & 4 & 3 \\ 2 & 6 & \\ 5 & & \end{array} , \quad \begin{array}{ccc} 1 & 5 & 3 \\ 2 & 6 & \\ 4 & & \end{array}$$
$$g_{A,B} = \quad \epsilon \; - \quad (4, 5) \; + \; (2, 4, 5) \; + \; (4, 6, 5) \; - \; (2, 4, 6, 5) \; + \; (2, 5)(4, 6)$$

The reader can verify that $g_{A,B} e_t = \mathbf{0}$, so that

$$e_t = e_{t_2} - e_{t_3} - e_{t_4} + e_{t_5} - e_{t_6}$$

where $t_2, \ldots, t_6$ are the second through sixth tableaux in the preceding list. Note that none of these arrays have the descent found in the second row of t.

Proposition 2.6.3 *Let t, A, and B, be as in the definition of a Garnir element. If $|A \cup B|$ is greater than the number of elements in column j of t, then $g_{A,B} e_t = \mathbf{0}$.*

Proof. First, we claim that

$$S^-_{A \cup B} e_t = 0 \tag{2.4}$$

Consider any $\sigma \in C_t$. By the hypothesis, there must be $a, b \in A \cup B$ such that a and b are in the same row of σt. But then $(a, b) \in S_{A \cup B}$ and $S^-_{A \cup B}\{\sigma t\} = 0$ by part 4 of the sign lemma (Lemma 2.4.1). Since this is true of every σ appearing in κ_t, the claim follows.

Now $S_{A \cup B} = \biguplus_\pi \pi(S_A \times S_B)$, so $S^-_{A \cup B} = g_{A,B}(S_A \times S_B)^-$. Substituting this into equation (2.4) yields

$$g_{A,B}(S_A \times S_B)^- e_t = 0 \tag{2.5}$$

and we need worry only about the contribution of $(S_A \times S_B)^-$. But we have $S_A \times S_B \subseteq C_t$. So if $\sigma \in S_A \times S_B$, then, by part 1 of the sign lemma,

$$\sigma^- e_t = \sigma^- C_t^-\{t\} = C_t^-\{t\} = e_t$$

Thus $(S_A \times S_B)^- e_t = |S_A \times S_B| e_t$, and dividing equation (2.5) by this cardinality yields the proposition. ∎

The reader may have noticed that when we eliminated the descent in row 2 of the preceding example, we introduced descents in some other places—e.g., in row 1 of t_3. Thus we need some measure of standardness that makes $t_2, \ldots, t_6$ closer to being standard than t. This is supplied by yet another partial order. Given t, consider its *column equivalence class*, or *column tabloid*,

$$[t] \stackrel{\text{def}}{=} C_t t$$

—i.e., the set of all tableaux obtained by rearranging elements within columns of t. We use vertical lines to denote the column tabloid, as in

$$\left| \begin{array}{c|c} 1 & 2 \\ 3 & \end{array} \right| = \left\{ \begin{array}{ccc} 1 & 2 & 3 \\ 3 & & 1 \end{array} , \begin{array}{c} 2 \\ 1 \end{array} \right\}$$

Replacing *row* by *column* everywhere in the definition of dominance for tabloids, we obtain a definition of column dominance for which we use the same symbol as for rows (the difference in the types of brackets used for the classes makes the necessary distinction).

Our proof that the standard polytabloids span S^λ follows Peel [Pee 75].

Theorem 2.6.4 *The set*

$$\{e_t \mid t \text{ is a standard } \lambda\text{-tableau}\} \tag{2.6}$$

spans S^λ.

First note that if e_t is in the span of the set (2.6), then so is e_s for any $s \in [t]$, by the remarks at the beginning of this section. Thus we may always take t to have increasing columns.

The poset of column tabloids has a maximum element $[t_0]$, where t_0 is obtained by numbering the cells of each column consecutively from top to bottom, starting with the leftmost column and working right. Since t_0 is standard, we are done for this equivalence class.

Now pick any tableau t. By induction, we may assume that every tableau s with $[s] \rhd [t]$ is in the span of (2.6). If t is standard, then we're done. If not, then there must be a descent in some row i (since columns increase). Let the columns involved be the jth and $(j+1)^{\text{st}}$ with entries $a_1 < a_2 < \cdots < a_p$ and $b_1 < b_2 < \cdots < b_q$, respectively. Thus we have the following situation in t:

$$
\begin{array}{cc}
a_1 & b_1 \\
& \wedge \\
a_2 & b_2 \\
& \wedge \\
\vdots & \vdots \\
& \wedge \\
a_i & > \quad b_i \\
\wedge & \\
\vdots & \vdots \\
\wedge & b_q \\
a_p &
\end{array}
$$

Take $A = \{a_i, \ldots, a_p\}$ and $B = \{b_1, \ldots, b_i\}$ with associated Garnir element $g_{A,B} = \sum_\pi (\operatorname{sgn} \pi)\pi$. By Proposition 2.6.3 we have $g_{A,B} e_t = \mathbf{0}$, so that

$$
e_t = -\sum_{\pi \neq \epsilon} (\operatorname{sgn} \pi) e_{\pi t} \tag{2.7}
$$

But $b_1 < \cdots < b_i < a_i < \cdots < a_p$ implies that $[\pi t] \unrhd [t]$ for $\pi \neq \epsilon$ by the column analog of the dominance lemma for tabloids (Lemma 2.5.5). Thus all terms on the right side of (2.7), and hence e_t itself, are in the span of the standard polytabloids. ∎

To summarize our results, let

$$
f^\lambda = \text{ the number of standard } \lambda\text{-tableaux}
$$

Then the following is true over any base field.

Theorem 2.6.5 *For any partition* λ

1. $\{e_t \mid t \text{ is a standard } \lambda\text{-tableau}\}$ *is a basis for* S^λ,

2. $\dim S^\lambda = f^\lambda$, *and*

3. $\sum_{\lambda \vdash n} (f^\lambda)^2 = n!$.

Proof. The first two parts are immediate. The third follows from the fact (Proposition 1.10.1) that, for any group G,

$$\sum_V (\dim V)^2 = |G|$$

where the sum is over all irreducible G-modules. $\blacksquare$

2.7 Young's Natural Representation

The matrices for the module S^λ in the standard basis form what is known as *Young's natural representation*. In this section we illustrate how these arrays are obtained.

Since $\mathcal{S}_n$ is generated by the transpositions $(k, k+1)$ for $k = 1, \ldots, n-1$, one need compute only the matrices corresponding to these group elements. If t is a standard tableau, we get the tth column of the matrix for $\pi \in \mathcal{S}_n$ by expressing $\pi e_t = e_{\pi t}$ as a sum of standard polytabloids. When $\pi = (k, k+1)$, there are three cases.

1. If k and $k+1$ are in the same column of t, then $(k, k+1) \in C_t$ and

$$(k, k+1)e_t = -e_t$$

2. If k and $k+1$ are in the same row of t, then $(k, k+1)t$ has a descent in that row. Applying the appropriate Garnir element,

$$(k, k+1)e_t = e_t \pm \text{ other polytabloids } e_{t'} \text{ such that } [t'] \rhd [t]$$

3. If k and $k+1$ are not in the same row or column of t, then the tableau $t' = (k, k+1)t$ is standard and

$$(k, k+1)e_t = e_{t'}$$

The e_t in the sum for case 2 comes from the term of equation (2.7) corresponding to $\pi = (k, k+1)$. Although we do not have an explicit expression for the rest of the terms, repeated application of 1 through 3 will compute them (by reverse induction on the column dominance ordering). This is the straightening algorithm mentioned at the beginning of Section 2.6.

By way of example, let us compute the matrices for the representation of $\mathcal{S}_3$ corresponding to $\lambda = (2, 1)$. The two standard tableaux of shape λ are

$$t_1 = \begin{matrix} 1 \ 3 \\ 2 \end{matrix} \quad \text{and} \quad t_2 = \begin{matrix} 1 \ 2 \\ 3 \end{matrix}$$

Note that we have chosen our indices so that $[t_1] \rhd [t_2]$. This makes the computations in case 2 easier.

For the transposition $(1,2)$ we have

$$(1,2)e_{t_1} = \begin{array}{|c|c|}\hline 2 & 3 \\\hline 1 \\\cline{1-1}\end{array} - \begin{array}{|c|c|}\hline 1 & 3 \\\hline 2 \\\cline{1-1}\end{array} = -e_{t_1}$$

as predicted by case 1. Since

$$(1,2)t_2 = \begin{array}{|c|c|}\hline 2 & 1 \\\hline 3 \\\cline{1-1}\end{array}$$

has a descent in row 1, we must find a Garnir element. Taking $A = \{2,3\}$ and $B = \{1\}$ gives tableaux

$$\begin{array}{|c|c|}\hline 2 & 1 \\\hline 3 \\\cline{1-1}\end{array} = (1,2)t_2, \qquad \begin{array}{|c|c|}\hline 1 & 2 \\\hline 3 \\\cline{1-1}\end{array} = t_2, \quad \text{and} \quad \begin{array}{|c|c|}\hline 1 & 3 \\\hline 2 \\\cline{1-1}\end{array} = t_1$$

with element

$$g_{A,B} = \epsilon - (1,2) + (1,3,2)$$

Thus

$$e_{(1,2)t_2} - e_{t_2} + e_{t_1} = 0$$

or

$$(1,2)e_{t_2} = e_{t_2} - e_{t_1}$$

(This result can be obtained more easily by merely noting that

$$(1,2)e_{t_2} = \begin{array}{|c|c|}\hline 1 & 2 \\\hline 3 \\\cline{1-1}\end{array} - \begin{array}{|c|c|}\hline 3 & 1 \\\hline 2 \\\cline{1-1}\end{array}$$

and expressing this as a linear combination of the e_{t_i}.) Putting everything together, we obtain the matrix

$$X(\,(1,2)\,) = \begin{pmatrix} -1 & -1 \\ 0 & 1 \end{pmatrix}$$

The transposition $(2,3)$ is even easier to deal with, since case 3 always applies, yielding

$$(2,3)t_1 = t_2$$

and

$$(2,3)t_2 = t_1$$

Thus

$$X(\,(2,3)\,) = \begin{pmatrix} 0 & 1 \\ 1 & 0 \end{pmatrix}$$

We have already seen these two matrices in Example 1.9.5. Since the adjacent transpositions generate $\mathcal{S}_n$, all other matrices must agree as well.

2.8 The Branching Rule

It is natural to ask what happens when we restrict or induce an irreducible representation S^λ of $\mathcal{S}_n$ to $\mathcal{S}_{n-1}$ or $\mathcal{S}_{n+1}$, respectively. The branching theorem gives a simple answer to that question.

Intuitively, these two operations correspond to either removing or adding a node to the Ferrers diagram for λ.

Definition 2.8.1 If λ is a diagram, then an *inner corner of* λ is a node $(i,j) \in \lambda$ whose removal leaves the Ferrers diagram of a partition. Any partition obtained by such a removal is denoted λ^-. An *outer corner of* λ is a node $(i,j) \notin \lambda$ whose addition produces the Ferrers diagram of a partition. Any partition obtained by such an addition is denoted λ^+. ∎

Note that the inner corners of λ are exactly those nodes at end of a row and column of λ. For example, if $\lambda = (5,4,4,2)$, then the inner corners are enlarged and the outer corners marked with open circles in the following diagram:

So, after removal, we could have

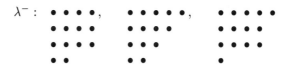

whereas after addition, the possibilities are

These are exactly the partitions that occur in restriction and induction. In particular,
$$S^{(5,4,4,2)} \downarrow_{\mathcal{S}_{14}} \cong S^{(4,4,4,2)} \oplus S^{(5,4,3,2)} \oplus S^{(5,4,4,1)}$$
and
$$S^{(5,4,4,2)} \uparrow^{\mathcal{S}_{16}} \cong S^{(6,4,4,2)} \oplus S^{(5,5,4,2)} \oplus S^{(5,4,4,3)} \oplus S^{(5,4,4,2,1)}$$

Before proving the branching rule itself, we need a result about dimensions.

Lemma 2.8.2 *We have*

$$f^\lambda = \sum_{\lambda^-} f^{\lambda^-}$$

Proof. Every standard tableau of shape $\lambda \vdash n$ consists of n in some inner corner together with a standard tableau of shape $\lambda^- \vdash n - 1$. The result follows. ∎

Theorem 2.8.3 (Branching Rule) *If $\lambda \vdash n$, then*

1. $S^\lambda \downarrow_{\mathcal{S}_{n-1}} \cong \bigoplus_{\lambda^-} S^{\lambda^-}$, *and*

2. $S^\lambda \uparrow^{\mathcal{S}_{n+1}} \cong \bigoplus_{\lambda^+} S^{\lambda^+}$.

Proof. [Pee 75] 1. Let the inner corners of λ appear in rows $r_1 < r_2 < \cdots < r_k$. For each i, let λ^i denote the partition λ^- obtained by removing the corner cell in row r_i. In addition, if n is at the end of row r_i of tableau t (respectively, in row r_i of tabloid $\{t\}$), then t^i (respectively, $\{t^i\}$) will be the array obtained by removing the n.

Now given any group G with module V and submodule W, it is easy to see that

$$V \cong W \oplus (V/W)$$

where V/W is the quotient space. (See Exercise 7 in Chapter 1.) Thus it suffices to find a chain of subspaces

$$\{0\} = V^{(0)} \subset V^{(1)} \subset V^{(2)} \subset \cdots \subset V^{(k)} = S^\lambda$$

such that $V^{(i)}/V^{(i-1)} \cong S^{\lambda^i}$ as $\mathcal{S}_{n-1}$-modules for $1 \le i \le k$. Let $V^{(i)}$ be the vector space spanned by the standard polytabloids e_t, where n appears in t at the end of one of rows r_1 through r_i. We show that the $V^{(i)}$ are our desired modules as follows.

Define maps $\theta_i : M^\lambda \to M^{\lambda^i}$ by linearly extending

$$\{t\} \xrightarrow{\theta_i} \begin{cases} \{t_i\} & \text{if } n \text{ is in row } r_i \text{ of } \{t\} \\ 0 & \text{otherwise} \end{cases}$$

The reader can quickly verify that θ_i is an $\mathcal{S}_{n-1}$-homomorphism. Furthermore, for standard t we have

$$e_t \xrightarrow{\theta_i} \begin{cases} e_{t^i} & \text{if } n \text{ is in row } r_i \text{ of } t \\ 0 & \text{if } n \text{ is in row } r_j \text{ of } t, \text{ where } j < i. \end{cases} \tag{2.8}$$

This is because any tabloid appearing in e_t, t standard, has n in the same row or higher than in t.

Since the standard polytabloids form a basis for the corresponding Specht module, the two parts of (2.8) show that

$$\theta_i V^{(i)} = S^{\lambda^i} \tag{2.9}$$

and

$$V^{(i-1)} \subseteq \ker \theta_i \qquad (2.10)$$

From equation (2.10), we can construct the chain

$$\{\mathbf{0}\} = V^{(0)} \subseteq V^{(1)} \cap \ker \theta_1 \subseteq V^{(1)} \subseteq V^{(2)} \cap \ker \theta_2 \subseteq V^{(2)} \subseteq \cdots \subseteq V^{(k)} = S^{\lambda} \qquad (2.11)$$

But from equation (2.9)

$$\dim \frac{V^{(i)}}{V^{(i)} \cap \ker \theta_i} = \dim \theta_i V^{(i)} = f^{\lambda^i}$$

By the preceding lemma, the dimensions of these quotients add up to $\dim S^{\lambda}$. Since this leaves no space to insert extra modules, the chain (2.11) must have equality for the first, third, etc. containments. Furthermore,

$$\frac{V^{(i)}}{V^{(i-1)}} \cong \frac{V^{(i)}}{V^{(i)} \cap \ker \theta_i} \cong S^{\lambda^i}$$

as desired. ■

2. We will show that this part follows from the first by Frobenius reciprocity (Theorem 1.12.6). In fact, parts 1 and 2 can be shown to be equivalent by the same method.

Let χ^{λ} be the character of S^{λ}. If $S^{\lambda} \uparrow^{S_{n+1}} \cong \oplus_{\mu \vdash n+1} m_{\mu} S^{\mu}$, then by taking characters, $\chi^{\lambda} \uparrow^{S_{n+1}} \cong \sum_{\mu \vdash n+1} m_{\mu} \chi^{\mu}$. The multiplicities are given by

$$
\begin{aligned}
m_{\mu} &= \langle \chi^{\lambda} \uparrow^{S_{n+1}}, \chi^{\mu} \rangle && \text{(Corollary 1.9.4, part 2)} \\
&= \langle \chi^{\lambda}, \chi^{\mu} \downarrow_{S_n} \rangle && \text{(Frobenius reciprocity)} \\
&= \langle \chi^{\lambda}, \sum_{\mu^-} \chi^{\mu^-} \rangle && \text{(branching rule, part 1)} \\
&= \begin{cases} 1 & \text{if } \lambda = \mu^- \\ 0 & \text{otherwise} \end{cases} && \text{(Corollary 1.9.4, part 4)} \\
&= \begin{cases} 1 & \text{if } \mu = \lambda^+ \\ 0 & \text{otherwise} \end{cases} && \text{(definition of } \mu^- \text{ and } \lambda^+)
\end{aligned}
$$

This finishes the proof. ■

2.9   The Decomposition of M^{μ}

We would like to know the multiplicity $m_{\lambda\mu}$ of the Specht module S^{λ} in M^{μ}. In fact, we can give a nice combinatorial description of these numbers in terms of tableaux that allow repetition of entries.

Definition 2.9.1 A *generalized Young tableau of shape* λ, T, is obtained by replacing the nodes of λ with positive integers, repetitions allowed. The *type* or *content of* T is the composition $\mu = (\mu_1, \mu_2, \ldots, \mu_m)$, where μ_i equals the number of i's in T. Let

$$T_{\lambda\mu} = \{T \mid T \text{ has shape } \lambda \text{ and content } \mu\} \quad \blacksquare$$

Note that capital letters will be used for generalized tableaux. One such array is

$$T = \begin{matrix} 4 & 1 & 4 \\ 1 & 3 & \end{matrix}$$

of shape $(3,2)$ and content $(2,0,1,2)$.

We will show that $\mathbf{C}[\mathcal{T}_{\lambda\mu}]$ is really a new copy of M^μ. For the rest of this section and the next:

fix a tableau t of shape λ and content (1^n)

In all our examples we will use $\lambda = (3,2)$ and

$$t = \begin{matrix} 1 & 2 & 3 \\ 4 & 5 & \end{matrix}$$

If T is any λ-tableau, then let $T(i)$ denote the element of T in the same position as the i in the fixed tableau t. With t and T as before,

$$T = \begin{matrix} T(1) & T(2) & T(3) \\ T(4) & T(5) & \end{matrix} \tag{2.12}$$

so that $T(1) = T(3) = 4$, $T(2) = T(4) = 1$, and $T(5) = 3$.

Now given any tabloid $\{s\}$ of shape μ, produce a tableau $T \in \mathcal{T}_{\lambda\mu}$ by letting

$$T(i) = \text{the number of the row in which } i \text{ appears in } \{s\}$$

For example, suppose $\mu = (2,2,1)$ and

$$\{s\} = \begin{array}{|c|c|} \hline 2 & 3 \\ \hline 1 & 5 \\ \hline 4 & \\ \hline \end{array}$$

Then the 2 and 3 are in row one of $\{s\}$, so the 2 and 3 of t get replaced by ones, etc., to obtain

$$T = \begin{matrix} 2 & 1 & 1 \\ 3 & 2 & \end{matrix}$$

Note that the shape of $\{s\}$ becomes the content of T, as desired. Also, it is clear that the map $\{s\} \overset{\theta}{\to} T$ is a bijection between bases for M^μ and $\mathbf{C}[\mathcal{T}_{\lambda\mu}]$, so they are isomorphic as vector spaces.

We now need to define an action of $\mathcal{S}_n$ on generalized tableaux so that $M^\mu \cong \mathbf{C}[\mathcal{T}_{\lambda\mu}]$ as modules. If $\pi \in \mathcal{S}_n$ and $T \in \mathcal{T}_{\lambda\mu}$, then we let πT be the tableau such that

$$(\pi T)(i) \overset{\text{def}}{=} T(\pi^{-1}i)$$

To illustrate with our canonical tableau (2.12):

$$(1,2,4)T = \begin{matrix} T(4) & T(1) & T(3) \\ T(2) & T(5) & \end{matrix}$$

In particular,

$$(1,2,4)\ \begin{array}{ccc} 2 & 1 & 1 \\ 3 & 2 & \end{array} \ = \ \begin{array}{ccc} 3 & 2 & 1 \\ 1 & 2 & \end{array}$$

Note that although $\pi \in S_n$ acts on the elements of $\{s\}$ (replacing i by πi), it acts on the places in T (moving $T(i)$ to the position of $T(\pi i)$). To see why this is the correct definition for making θ into a module isomorphism, consider $\theta(\{s\}) = T$ and $\pi \in S_n$. Thus we want $\theta(\pi\{s\}) = \pi T$. But this forces us to set

$$\begin{aligned} (\pi T)(i) &= \text{row number of } i \text{ in } \pi\{s\} \\ &= \text{row number of } \pi^{-1}i \text{ in } \{s\} \\ &= T(\pi^{-1}i) \end{aligned}$$

We have proved, by fiat, the following.

Proposition 2.9.2 *For any given partition λ, the modules M^μ and $\mathbf{C}[\mathcal{T}_{\lambda\mu}]$ are isomorphic.* ∎

Recall (Proposition 1.7.10) that the multiplicity of S^λ in M^μ is given by $\dim \operatorname{Hom}(S^\lambda, M^\mu)$. We will first construct certain homomorphisms from M^λ to M^μ using our description of the latter in terms of generalized tableaux and then restrict to S^λ. The row and column equivalence classes of a generalized tableau T, denoted $\{T\}$ and $[T]$, respectively, are defined in the obvious way. Let $\{t\} \in M^\lambda$ be the tabloid associated with our fixed tableau.

Definition 2.9.3 For each $T \in \mathcal{T}_{\lambda\mu}$, the *homomorphism corresponding to T* is the map $\theta_T \in \operatorname{Hom}(M^\lambda, M^\mu)$ given by

$$\{t\} \overset{\theta_T}{\to} \sum_{S \in \{T\}} S$$

and extension using the cyclicity of M^λ. Note that θ_T is actually a homomorphism into $\mathbf{C}[\mathcal{T}_{\lambda\mu}]$, but that should cause no problems in view of the previous proposition. ∎

Extension by cyclicity means that, since every element of M^λ is of the form $g\{t\}$ for some $g \in \mathbf{C}[S_n]$, we must have $\theta_T(g\{t\}) = g\sum_{S \in \{T\}} S$ (that θ_T respects the group action and linearity). For example, if

$$T = \begin{array}{ccc} 2 & 1 & 1 \\ 3 & 2 & \end{array}$$

then

$$\theta_T\{t\} = \begin{array}{cc} 2\ 1\ 1 \\ 3\ 2 \end{array} + \begin{array}{cc} 1\ 2\ 1 \\ 3\ 2 \end{array} + \begin{array}{cc} 1\ 1\ 2 \\ 3\ 2 \end{array} + \begin{array}{cc} 2\ 1\ 1 \\ 2\ 3 \end{array} + \begin{array}{cc} 1\ 2\ 1 \\ 2\ 3 \end{array} + \begin{array}{cc} 1\ 1\ 2 \\ 2\ 3 \end{array}$$

and

$$\theta_T(1,2,4)\{t\} = \begin{array}{cc} 3\ 2\ 1 \\ 1\ 2 \end{array} + \begin{array}{cc} 3\ 1\ 1 \\ 2\ 2 \end{array} + \begin{array}{cc} 3\ 1\ 2 \\ 1\ 2 \end{array} + \begin{array}{cc} 2\ 2\ 1 \\ 1\ 3 \end{array} + \begin{array}{cc} 2\ 1\ 1 \\ 2\ 3 \end{array} + \begin{array}{cc} 2\ 1\ 2 \\ 1\ 3 \end{array}$$

Now we obtain elements of $\mathrm{Hom}(S^\lambda, M^\mu)$ by letting

$$\bar{\theta}_T = \text{ the restriction of } \theta_T \text{ to } S^\lambda$$

If t is our fixed tableau, then

$$\bar{\theta}_T(\boldsymbol{e}_t) = \bar{\theta}_T(\kappa_t\{\boldsymbol{t}\}) = \kappa_t(\theta_T\{\boldsymbol{t}\}) = \kappa_t\Big(\sum_{S\in\{T\}} \boldsymbol{S}\Big)$$

This last expression could turn out to be zero (thus forcing $\bar{\theta}_T$ to be the zero map by the cyclicity S^λ) because of the following.

Proposition 2.9.4 *If t is the fixed λ-tableau and $T \in \mathcal{T}_{\lambda\mu}$, then $\kappa_t\boldsymbol{T} = \boldsymbol{0}$ if and only if T has two equal elements in some column.*

Proof. If $\kappa_t\boldsymbol{T} = \boldsymbol{0}$, then

$$\boldsymbol{T} + \sum_{\substack{\pi\in C_t \\ \pi\neq\epsilon}} (\mathrm{sgn}\,\pi)\pi\boldsymbol{T} = \boldsymbol{0}$$

So we must have $\boldsymbol{T} = \pi\boldsymbol{T}$ for some $\pi \in C_t$ with $\mathrm{sgn}\,\pi = -1$. But then the elements corresponding to any nontrivial cycle of π are all equal and in the same column.

Now suppose that $T(i) = T(j)$ are in the same column of T. Then

$$(\epsilon - (i,j))\boldsymbol{T} = \boldsymbol{0}$$

But $\epsilon - (i,j)$ is a factor of κ_t by part 3 of the sign lemma (Lemma 2.4.1), so $\kappa_t\boldsymbol{T} = \boldsymbol{0}$. ∎

In light of the previous proposition, we can eliminate possibly trivial $\bar{\theta}_T$ from consideration by concentrating on the analog of standard tableaux for arrays with repetitions.

Definition 2.9.5 A generalized tableau is *semistandard* if its rows weakly increase and its columns strictly increase. We let $\mathcal{T}^0_{\lambda\mu}$ denote the set of semistandard λ-tableau of type μ. ∎

The tableau

$$T = \begin{array}{ccc} 1 & 1 & 2 \\ 2 & 3 & \end{array}$$

is semistandard, whereas

$$T = \begin{array}{ccc} 2 & 1 & 1 \\ 3 & 2 & \end{array}$$

is not. The homomorphisms corresponding to semistandard tableaux are the ones we have been looking for. Specifically, we will show that they form a basis for $\mathrm{Hom}(S^\lambda, M^\mu)$.

2.10 The Semistandard Basis for $\text{Hom}(S^\lambda, M^\mu)$

This section is devoted to proving the following theorem.

Theorem 2.10.1 *The set*

$$\{\bar{\theta}_T \mid T \in \mathcal{T}^0_{\lambda\mu}\}$$

is a basis for $\text{Hom}(S^\lambda, M^\mu)$.

In many ways the proof will parallel the one given to show that the standard polytabloids form a basis for S^λ in Sections 2.5 and 2.6.

As usual, we need appropriate partial orders. The dominance and column dominance orderings for generalized tableaux are defined in exactly the same way as for tableaux without repetitions (see Definition 2.5.4). For example, if

$$[S] = \begin{array}{|c|c|c|} \hline 2 & 1 & 1 \\ \hline 3 & 2 \\ \hline \end{array} \quad \text{and} \quad [T] = \begin{array}{|c|c|c|} \hline 1 & 1 & 2 \\ \hline 2 & 3 \\ \hline \end{array}$$

then $[S]$ corresponds to the sequence of compositions

$$\lambda^1 = (0, 1, 1), \lambda^2 = (1, 2, 1), \lambda^3 = (2, 2, 1)$$

whereas $[T]$ has

$$\mu^1 = (1, 1, 0), \mu^2 = (2, 1, 1), \mu^3 = (2, 2, 1)$$

Since $\lambda^i \trianglelefteq \mu^i$ for all i, $[S] \trianglelefteq [T]$. The dominance lemma for tabloids (Lemma 2.5.5) and its corollary (Corollary 2.5.6) have analogs in this setting. Their proofs, being similar, are omitted.

Lemma 2.10.2 (Dominance Lemma for Generalized Tableaux) *Suppose $k < l$ and k is in a column to the left of l in T. Then*

$$[T] \rhd [S]$$

where S is the tableau obtained by interchanging k and l in T. ∎

Corollary 2.10.3 *If T is semistandard and $S \in \{T\}$ is different from T, then*

$$[T] \rhd [S] \quad ∎$$

Thus, if T is semistandard, then $[T]$ is the largest equivalence class to appear in $\theta_T\{t\}$.

Before proving independence of the $\bar{\theta}_T$, we must cite some general facts about vector spaces. Let V be a vector space and pick out a fixed basis $\mathcal{B} = \{b_1, b_2, \ldots, b_n\}$. If $v \in V$, then we say that b_i *appears in* v if $v = \sum_i c_i b_i$ with $c_i \neq 0$. Suppose that V is endowed with an equivalence relation whose equivalence classes are denoted $[v]$, and with a partial order on these classes. We can generalize Lemma 2.5.8 as follows.

Lemma 2.10.4 *Let V and $\mathcal{B}$ be as before and consider a set of vectors $\mathbf{v}_1, \mathbf{v}_2, \ldots, \mathbf{v}_m \in V$. Suppose that, for all i, there exists a $\mathbf{b}_i \in \mathcal{B}$ appearing in $\mathbf{v}_i$ such that*

1. *$[\mathbf{b}_i] \unrhd [\mathbf{b}]$ for every $\mathbf{b} \neq \mathbf{b}_i$ appearing in $\mathbf{v}_i$, and*

2. *the $[\mathbf{b}_i]$ are all distinct.*

Then the $\mathbf{v}_i$ are linearly independent. ■

We also need a simple lemma about independence of linear transformations.

Lemma 2.10.5 *Let V and W be vector spaces and let $\theta_1, \theta_2, \ldots, \theta_m$ be linear maps from V to W. If there exists a $\mathbf{v} \in V$ such that $\theta_1(\mathbf{v}), \theta_2(\mathbf{v}), \ldots, \theta_m(\mathbf{v})$ are independent in W, then the θ_i are independent as linear transformations.*

Proof. Suppose not. Then there are constants c_i, not all zero, such that $\sum_i c_i \theta_i$ is the zero map. But then $\sum_i c_i \theta_i(\mathbf{v}) = \mathbf{0}$ for all $\mathbf{v} \in V$, a contradiction to the hypothesis of the lemma. ■

Proposition 2.10.6 *The set*

$$\{\bar{\theta}_T \mid T \in \mathcal{T}^0_{\lambda\mu}\}$$

is independent.

Proof. Let $T_1, T_2, \ldots, T_m$ be the elements of $\mathcal{T}_{\lambda\mu}$. By the previous lemma, it suffices to show that $\bar{\theta}_{T_1} e_t, \bar{\theta}_{T_2} e_t, \ldots, \bar{\theta}_{T_m} e_t$ are independent, where t is our fixed tableau. For all i we have

$$\bar{\theta}_{T_i} e_t = \theta_{T_i} \kappa_t \{t\} = \kappa_t \theta_{T_i} \{t\}$$

Now T_i is semistandard, so $[T_i] \rhd [S]$ for any other summand S in $\theta_{T_i}\{t\}$ (Corollary 2.10.3). The same is true for summands of $\kappa_t \theta_{T_i}\{t\}$, since the permutations in κ_t don't change the column equivalence class. Also the $[T_i]$ are all distinct, since no equivalence class has more than one semistandard tableau. Hence the $\kappa_t \theta_{T_i}\{t\} = \bar{\theta}_{T_i} e_t$ satisfy the hypotheses of Lemma 2.10.4, making them independent. ■

To prove that the $\bar{\theta}_T$ span, we need a lemma.

Lemma 2.10.7 *Consider $\theta \in \mathrm{Hom}(S^\lambda, M^\mu)$. Write*

$$\theta e_t = \sum_T c_T T$$

where t is the fixed tableau of shape λ.

1. *If $\pi \in C_t$ and $T_1 = \pi T_2$, then $c_{T_1} = (\mathrm{sgn}\,\pi)c_{T_2}$.*

2. *Every T_1 with a repetition in some column has $c_{T_1} = 0$.*

3. *If θ is not the zero map, then there exists a semistandard T_2 having $c_{T_2} \neq 0$.*

Proof. 1. Since $\pi \in C_t$, we have

$$\pi(\theta e_t) = \theta(\pi \kappa_t\{t\}) = \theta((\operatorname{sgn}\pi)\kappa_t\{t\}) = (\operatorname{sgn}\pi)(\theta e_t)$$

Therefore, $\theta e_t = \sum_T c_T T$ implies

$$\pi \sum_T c_T T = \pi(\theta e_t) = (\operatorname{sgn}\pi)(\theta e_t) = (\operatorname{sgn}\pi) \sum_T c_T T$$

Comparing coefficients of πT_2 on the left and T_1 on the right yields $c_{T_2} = (\operatorname{sgn}\pi)c_{T_1}$, which is equivalent to part 1.

2. By hypothesis, there exists $(i,j) \in C_t$ with $(i,j)T_1 = T_1$. But then $c_{T_1} = -c_{T_1}$ by what we just proved, forcing this coefficient to be zero.

3. Since $\theta \neq 0$ we can pick T_2 with $c_{T_2} \neq 0$ such that $[T_2]$ is maximal. We claim that T_2 can be taken to be semistandard. By parts 1 and 2, we can choose T_2 so that its columns strictly increase.

Suppose, toward a contradiction, that we have a descent in row i. Thus T_2 has a pair of columns that look like

$$
\begin{array}{ccc}
 & b_1 & \\
 & \wedge & \\
 & b_2 & \\
 & \wedge & \\
 & \vdots & \\
 & \wedge & \\
a_i & > & b_i \\
\wedge & & \\
\vdots & & \\
\wedge & & \\
a_p & &
\end{array}
$$

Choose A and B as usual, and let $g_{A,B} = \sum_\pi (\operatorname{sgn}\pi)\pi$ be the associated Garnir element. We have

$$g_{A,B}\left(\sum_T c_T T\right) = g_{A,B}(\theta e_t) = \theta(g_{A,B}e_t) = \theta(0) = 0$$

Now T_2 appears in $g_{A,B}T_2$ with coefficient 1 (since the permutations in $g_{A,B}$ move distinct elements of T_2). So to cancel T_2 in the previous equation, there must be a $T \neq T_2$ with $\pi T = T_2$ for some π in $g_{A,B}$. Thus T is just T_2 with some of the a_j's and b_k's exchanged. But then $[T] \rhd [T_2]$ by the dominance lemma for generalized tableaux (Lemma 2.10.2). This contradicts our choice of T_2. ∎

We are now in a position to prove that the $\overline{\theta}_T$ generate $\operatorname{Hom}(S^\lambda, M^\mu)$.

Proposition 2.10.8 *The set*

$$\{\bar{\theta}_T \mid T \in T^0_{\lambda\mu}\}$$

spans $\mathrm{Hom}(S^\lambda, M^\mu)$.

Proof. Pick any $\theta \in \mathrm{Hom}(S^\lambda, M^\mu)$ and write

$$\theta e_t = \sum_T c_T T \qquad (2.13)$$

Consider

$$L_\theta = \{S \in T^0_{\lambda\mu} \mid [S] \trianglelefteq [T] \text{ for some } T \text{ appearing in } \theta e_t\}$$

In poset terminology, L_θ corresponds to the *lower order ideal* generated by the T in θe_t. We prove this proposition by induction on $|L_\theta|$.

If $|L_\theta| = 0$, then θ is the zero map by part 3 of the previous lemma. Such a θ is surely generated by our set!

If $|L_\theta| > 0$, then in equation (2.13) we can find a semistandard T_2 with $c_{T_2} \neq 0$. Furthermore, it follows from the proof of part 3 in Lemma 2.10.7 that we can choose $[T_2]$ maximal among those tableaux that appear in the sum. Now consider

$$\theta_2 = \theta - c_{T_2} \bar{\theta}_{T_2}$$

We claim that L_{θ_2} is a subset of L_θ with T_2 removed. First of all, every S appearing in $\bar{\theta}_{T_2} e_t$ satisfies $[S] \trianglelefteq [T_2]$ (see the comment after Corollary 2.10.3) so $L_{\theta_2} \subseteq L_\theta$. Furthermore, by part 1 of Lemma 2.10.7, every S with $[S] = [T_2]$ appears with the same coefficient in θe_t and $c_{T_2} \bar{\theta}_{T_2} e_t$. Thus $T_2 \notin L_{\theta_2}$, since $[T_2]$ is maximal. By induction, θ_2 is in the span of the $\bar{\theta}_T$ and thus θ is as well.

This completes the proof of the proposition and of Theorem 2.10.1. ∎

2.11 Kostka Numbers and Young's Rule

The Kostka numbers count semistandard tableaux.

Definition 2.11.1 The *Kostka numbers* are

$$K_{\lambda\mu} = |T^0_{\lambda\mu}| \quad \blacksquare$$

As an immediate corollary of the semistandard basis theorem (Theorem 2.10.1), we have Young's rule.

Theorem 2.11.2 (Young's Rule) *The multiplicity of S^λ in M^μ is equal to the number of semistandard tableaux of shape λ and content μ, i.e.,*

$$M^\mu \cong \bigoplus_\lambda K_{\lambda\mu} S^\lambda \quad \blacksquare$$

Note that by Corollary 2.4.7, we can restrict this direct sum to $\lambda \trianglerighteq \mu$. Let's look at some examples.

Example 2.11.3 Suppose $\mu = (2, 2, 1)$. Then the possible $\lambda \trianglerighteq \mu$ and the associated λ-tableaux of type μ are as follows.

$$\lambda^1 = (2,2,1), \quad \lambda^2 = (3,1,1), \quad \lambda^3 = (3,2), \quad \lambda^4 = (4,1), \quad \lambda^5 = (5)$$

$$
\begin{array}{ccccc}
\bullet\ \bullet & \bullet\ \bullet\ \bullet & \bullet\ \bullet\ \bullet & \bullet\ \bullet\ \bullet\ \bullet & \bullet\ \bullet\ \bullet\ \bullet\ \bullet \\
= \bullet\ \bullet & = \bullet & = \bullet\ \bullet & = \bullet & = \\
\bullet & \bullet & & &
\end{array}
$$

$$
\begin{array}{ccccc}
T:\ 1\ 1 & 1\ 1\ 2 & 1\ 1\ 2 & 1\ 1\ 2\ 2 & 1\ 1\ 2\ 2\ 3 \\
\ \ \ \ 2\ 2 & 2 & 2\ 3 & 3 & \\
\ \ \ \ 3 & 3 & & &
\end{array}
$$

$$
\begin{array}{ccccc}
 & & 1\ 1\ 3 & 1\ 1\ 2\ 3 & \\
 & & 2\ 2 & 2 &
\end{array}
$$

Thus

$$M^{(2,2,1)} \cong S^{(2,2,1)} \oplus S^{(3,3,1)} \oplus 2S^{(3,2)} \oplus 2S^{(4,1)} \oplus S^{(5)}$$

Example 2.11.4 For any μ, $K_{\mu\mu} = 1$. This is because the only μ-tableau of content μ is the one with all the 1s in row 1, all the 2s in row 2, etc. Of course, we saw this result in Corollary 2.4.7.

Example 2.11.5 For any μ, $K_{(n)\mu} = 1$. Obviously there's only one way to arrange a collection of numbers in weakly increasing order. It is also easy to see from a representation-theoretic viewpoint that M^μ must contain exactly one copy of $S^{(n)}$ (see Exercise 5).

Example 2.11.6 For any λ, $K_{\lambda(1^n)} = f^\lambda$ (the number of standard tableaux of shape λ). This says that

$$M^{(1^n)} \cong \bigoplus_\lambda f^\lambda S^\lambda$$

But $M^{(1^n)}$ is just the regular representation (Example 2.1.7) and $f^\lambda = \dim S^\lambda$ (Theorem 2.5.2). Thus this is merely the special case of Proposition 1.10.1, part 1, where $G = S_n$.

2.12 Exercises

1. Let ϕ^λ be the character of M^λ. Find (with proof) a formula for ϕ^λ_λ, the value of ϕ^λ on the conjugacy class K_λ.

2. Verify the details in Theorem 2.1.12.

3. Let $\lambda = (\lambda_1, \lambda_2, \ldots, \lambda_l)$ and $\mu = (\mu_1, \mu_2, \ldots, \mu_m)$ be partitions. Characterize the fact that λ is covered by μ if the ordering used is

   a. lexicographic,
   b. dominance.

4. Consider $S^{(n-1,1)}$, where each tabloid is identified with the element in its second row. Prove the following facts about this module and its character.

   a. We have

   $$S^{(n-1,1)} = \{c_1\mathbf{1} + c_2\mathbf{2} + \cdots + c_n\mathbf{n} \mid c_1 + c_2 + \cdots + c_n = 0\}$$

   b. For any $\pi \in \mathcal{S}_n$,

   $$\chi^{(n-1,1)}(\pi) = (\text{number of fixed-points of } \pi) - 1$$

5. Let the group G act on the set S. We say that G is *transitive* if, given any $s, t \in S$, there is a $g \in G$ with $gs = t$. The group is *doubly transitive* if, given any $s, t, u, v \in S$, there is a $g \in G$ with $gs = t$ and $gu = v$. Show the following.

   a. The orbits of G's action partition S.

   b. The multiplicity of the trivial representation in $V = \mathbf{C}[S]$ is the number of orbits. Thus if G is transitive, then the trivial representation occurs exactly once. What does this say about the module M^μ?

   c. If G is doubly transitive and V has character χ, then $\chi - 1$ is an irreducible character of G. *Hint:* Fix $s \in S$ and use Frobenius reciprocity on the stabilizer $G_s \leq G$.

   d. Use part (c) to conclude that in $\mathcal{S}_n$ the function

   $$f(\pi) = (\text{number of fixed-points of } \pi) - 1$$

   is an irreducible character.

6. Show that every irreducible character of $\mathcal{S}_n$ is an integer-valued function.

7. Define a lexicographic order on tabloids as follows. Associate with any $\{t\}$ the composition $\lambda = (\lambda_1, \lambda_2, \ldots, \lambda_n)$, where λ_i is the number of the row containing $n - i + 1$. If $\{s\}$ and $\{t\}$ have associated compositions λ and μ, respectively, then $\{s\} \leq \{t\}$ in *lexicographic order* (also called *last letter order*) if $\lambda \leq \mu$.

   a. Show that $\{s\} \trianglelefteq \{t\}$ implies $\{s\} \leq \{t\}$.

   b. Repeat Exercise 3 with tabloids in place of partitions.

8. Verify that the permutations π chosen after Definition 2.6.1 do indeed form a transversal for $\mathcal{S}_A \times \mathcal{S}_B$ in $\mathcal{S}_{A \cup B}$.

9. Verify the statements made in case 2 for the computation of Young's natural representation (page 74).

10. In $\mathcal{S}_n$ consider the transpositions $\tau_k = (k, k+1)$ for $k = 1, \ldots, n-1$.

    a. Prove that the τ_k generate $\mathcal{S}_n$ subject to the Coxeter relations

    $$
    \begin{array}{ll}
    \tau_k^2 = \epsilon & k = 1, \ldots, n-1 \\
    \tau_k \tau_{k+1} \tau_k = \tau_{k+1} \tau_k \tau_{k+1} & k = 1, \ldots, n-2 \\
    \tau_k \tau_l = \tau_l \tau_k & k, l = 1, \ldots, n-1 \text{ and } |k-l| \geq 2
    \end{array}
    $$

    b. Show that if G_n is a group generated by g_k for $k = 1, \ldots, n-1$ subject to the relations above (replacing τ_k by g_k) then $G_n \cong \mathcal{S}_n$. *Hint:* Induct on n using cosets of the subgroup generated by $g_1, \ldots, g_{n-2}$.

11. Fix a partition λ and fix an ordering of standard λ-tableaux $t_1, t_2, \ldots$. Define the *axial distance* from k to $k+1$ in tableau t_i to be

    $$
    \delta_i = \delta_i(k, k+1) = (c' - r') - (c - r)
    $$

    where c, c' and r, r' are the column and row coordinates of k and $k+1$, respectively, in t_i. *Young's seminormal form* assigns to each transposition $\tau = (k, k+1)$ the matrix $\rho_\lambda(\tau)$ with entries

    $$
    \rho_\lambda(\tau)_{i,j} = \begin{cases}
    1/\delta_i & \text{if } i = j \\
    1 - 1/\delta_i^2 & \text{if } \tau t_i = t_j \text{ and } i < j \\
    1 & \text{if } \tau t_i = t_j \text{ and } i > j \\
    0 & \text{otherwise}
    \end{cases}
    $$

    a. Show that every row and column of $\rho_\lambda(\tau)$ has at most two nonzero entries.

    b. Show that ρ_λ can be extended to a representation of $\mathcal{S}_n$, where $\lambda \vdash n$, by using the Coxeter relations of Exercise 10.

    c. Show that this representation is equivalent to the one afforded by S^λ.

12. All matrices for this problem have rows and columns indexed by partitions of n in dual lexicographic order. Define

    $$
    A = (|\mathcal{S}_\lambda \cap K_\mu|) \text{ and } B = (|\mathcal{S}_\mu| \cdot K_{\lambda\mu})
    $$

    (Be sure to distinguish between the conjugacy class K_μ and the Kostka number $K_{\lambda\mu}$.) Show that A, B are upper triangular matrices and that $C = B(A^t)^{-1}$ is the character table of $\mathcal{S}_n$ (with the columns listed in reverse order). *Hint:* Use Frobenius reciprocity.

    Use this method to calculate the character table for $\mathcal{S}_4$.

13. Prove in two ways that, up to sign, the determinant of the character table for $\mathcal{S}_n$ is

$$\prod_{\lambda \vdash n} \prod_{\lambda_i \in \lambda} \lambda_i$$

14. Prove the following results in two ways: once using representations and once combinatorially.

    a. If $K_{\lambda\mu} \neq 0$, then $\lambda \trianglerighteq \mu$.

    b. Suppose μ and ν are compositions with the same parts (only rearranged), then for any λ, $K_{\lambda\mu} = K_{\lambda\nu}$. *Hint:* For the combinatorial proof, consider the case where μ and ν differ by an adjacent transposition of parts.

15. Let G be a group and let $H \leq G$ have index two. Prove the following.

    a. H is normal in G.

    b. Every conjugacy class of G having nonempty intersection with H becomes a conjugacy class of H or splits into two conjugacy classes of H having equal size. Furthermore, the conjugacy class K of G does not split in H if and only if some $k \in K$ commutes with some $g \notin H$.

    c. Let χ be an irreducible character of G. Then $\chi{\downarrow}_H$ is irreducible or is the sum of two inequivalent irreducibles. Furthermore, $\chi{\downarrow}_H$ is irreducible if and only if $\chi(g) \neq 0$ for some $g \notin H$.

16. Let A_n denote the alternating subgroup of $\mathcal{S}_n$ and consider $\pi \in \mathcal{S}_n$ having cycle type $\lambda = (\lambda_1, \lambda_2, \ldots, \lambda_l)$.

    a. Show that $\pi \in A_n$ if and only if $n - l$ is even.

    b. Prove that the conjugacy classes of $\mathcal{S}_n$ that split in A_n are those where all parts of λ are odd and distinct.

17. Use the previous two exercises and the character table of $\mathcal{S}_4$ to find the character table of A_4.

Chapter 3

Combinatorial Algorithms

Many results about representations of the symmetric group can be approached in a purely combinatorial manner. The crucial link between these two viewpoints is the fact (Theorem 2.6.5, part 2) that the number of standard Young tableaux of given shape is the degree of the corresponding representation.

In the first two sections of this chapter we give two formulae for

$$f^\lambda = \dim S^\lambda$$

One is in terms of products, whereas the other involves determinants. Next we introduce the famed Robinson-Schensted algorithm [Rob 38, Sch 61], which provides a bijective proof of the identity

$$\sum_{\lambda \vdash n} (f^\lambda)^2 = n!$$

from part 3 of Theorem 2.6.5. This procedure has many surprising properties that are surveyed in the rest of the chapter. Included is a discussion of Schützenberger's jeu de taquin [Scü 76].

3.1 The Hook Formula

There is an amazingly simple product formula for f^λ, the number of standard λ-tableaux. It involves objects called hooks.

Definition 3.1.1 If $v = (i, j)$ is a node in the diagram of λ, then it has *hook*

$$H_v = H_{i,j} = \{(i, j') \mid j' \geq j\} \cup \{(i', j) \mid i' \geq i\}$$

with corresponding *hooklength*

$$h_v = h_{i,j} = |H_{i,j}| \ \blacksquare$$

To illustrate, if $\lambda = (4^2, 3^3, 1)$, then the dotted cells in

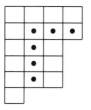

are the hook $H_{2,2}$ with hooklength $h_{2,2} = 6$.

It is now easy to state the hook formula of Frame, Robinson, and Thrall.

Theorem 3.1.2 (Hook Formula [FRT 54]) *If* $\lambda \vdash n$, *then*

$$f^\lambda = \frac{n!}{\prod_{(i,j) \in \lambda} h_{i,j}}$$

Before proving this theorem, let us pause for an example and an anecdote. Suppose we wish to calculate the number of standard Young tableaux of shape $\lambda = (2, 2, 1) \vdash 5$. The hooklengths are given in the array

4	2
3	1
1	

where $h_{i,j}$ is placed in cell (i, j). Thus

$$f^{(2,2,1)} = \frac{5!}{4 \cdot 3 \cdot 2 \cdot 1^2} = 5$$

This result can be verified by listing all possible tableaux:

$$
\begin{array}{ccccc}
1\,2, & 1\,2, & 1\,3, & 1\,3, & 1\,4 \\
3\,4 & 3\,5 & 2\,4 & 2\,5 & 2\,5 \\
5 & 4 & 5 & 4 & 3
\end{array}
$$

The tale of how the hook formula was born is an amusing one. One Thursday in May of 1953, Robinson was visiting Frame at Michigan State University. Discussing the work of Staal [Sta 50] (a student of Robinson), Frame was led to conjecture the hook formula. At first Robinson could not believe that such a simple formula existed, but after trying some examples he became convinced and together they proved the identity. On Saturday they went to the University of Michigan, where Frame presented their new result after a lecture by Robinson. This surprised Thrall, who was in the audience, because he had just proved the same result on the same day!

We will now demonstrate the hook formula using a probabilistic approach due to Greene, Nijenhuis, and Wilf [GNW 79]. It has the nice feature that the hooks play a very natural role in the proof. As a bonus, an algorithm is provided for generating a standard tableau of given shape at random.

Fix a partition $\lambda \vdash n$. If we can find a procedure that produces any given standard λ-tableau P with probability

$$\text{prob}(P) = \frac{\prod h_{i,j}}{n!}$$

then we will be done, since the distribution is uniform. Note that we will use capital letters such as P and Q to stand for *partial tableaux*—i.e., arrays with distinct entries whose rows and columns increase. Thus a partial tableau will be standard if its elements are precisely $\{1, 2, \ldots, n\}$.

Our candidate for the desired algorithm is as follows (where := is the PASCAL language replacement symbol).

GNW1 Pick a node $v \in \lambda$ with probability $1/n$.

GNW2 **While** v is not an inner corner **do**

> **begin**
>
> GNWa Pick a node $\bar{v} \in H_v - \{v\}$ with probability $1/(h_v - 1)$.
> GNWb $v := \bar{v}$.
>
> **end**

GNW3 Give the label n to the corner cell v that you've reached.

GNW4 Go back to step GNW1 with $\lambda := \lambda - \{v\}$ and $n := n - 1$, repeating this outer loop until all cells of λ are labeled.

The sequence of nodes generated by one pass through GNW1-3 is called a *trial*. We use the notation $v_1, v_2, \ldots$ for the cells of a trial with w being the final cell. As an example, a typical trial with probabilities is

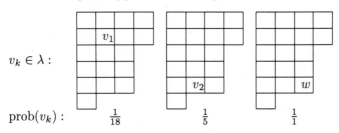

$$v_k \in \lambda:$$

$$\text{prob}(v_k): \qquad \frac{1}{18} \qquad \frac{1}{5} \qquad \frac{1}{1}$$

We must prove the following.

Proposition 3.1.3 ([GNW 79]) *Given a shape $\lambda \vdash n$, then GNW1-4 produces any particular standard tableau P of shape λ with*

$$\text{prob}(P) = \frac{\prod_{v \in \lambda} h_v}{n!}$$

Proof. It is clear that our algorithm produces a standard labeling of λ. To show that all labelings have the same probability, we induct on n.

Fix a λ-tableau P. Let w be the cell containing n and let $\overline{P}$ be P with the n deleted. Say $\overline{P}$ has shape $\overline{\lambda}$. Thus

$$\text{prob}(P) = \text{prob}(w)\,\text{prob}(\overline{P})$$

where prob(w) is the probability that a trial ends at w. By induction, it suffices to show that

$$\text{prob}(w) \stackrel{?}{=} \frac{\prod_{v \in \lambda} h_v/n!}{\prod_{\overline{v} \in \overline{\lambda}} h_{\overline{v}}/(n-1)!} \tag{3.1}$$

Everything cancels in this expression except for the terms corresponding to the following subset of λ:

$$W \;=\; \{v \neq w \mid w \in H_v,\}$$
$$=\; \{v \neq w \mid v \text{ is in the same row or column as } w\}$$

Hooklengths in W are all decreased by 1, so we can write equation (3.1) as

$$\text{prob}(w) \quad \stackrel{?}{=} \quad \frac{1}{n} \prod_{v \in W} \frac{h_v}{h_v - 1}$$
$$=\; \frac{1}{n} \prod_{v \in W} (1 + \frac{1}{h_v - 1})$$

Let w have coordinates (α, β). For convenience of notation, let $a_i = h_{i,\beta} - 1$ and $b_j = h_{\alpha,j} - 1$. Then we can rewrite the preceding fraction as

$$\text{prob}(w) = \text{prob}(\alpha, \beta) \stackrel{?}{=} \frac{1}{n} \prod_{i=1}^{\alpha-1} (1 + \frac{1}{a_i}) \prod_{j=1}^{\beta-1} (1 + \frac{1}{b_j}) \tag{3.2}$$

We can give a combinatorial interpretation to each term in the expansion of (3.2) as follows. Given a trial ending at (α, β), we let the *horizontal projection of the trial* be

$$I = \{i \neq \alpha \mid v = (i,j) \text{ for some } v \text{ on the trial}\}$$

The *vertical projection* is defined analogously. Let $\text{prob}_{I,J}(\alpha, \beta)$ denote the sum of the probabilities of all trials terminating at (α, β) with horizontal projection I and vertical projection J. So $\text{prob}(\alpha, \beta) = \sum_{I,J} \text{prob}_{I,J}(\alpha, \beta)$.

By way of illustration, let w and W be as shown in the diagram:

		W
		W
		W
		W
W	W	w

so $w = (\alpha, \beta) = (5,3)$. If $I = \{2,3\}$ and $J = \{2\}$, then the possible trials are

$$v_1 = (2,2), \quad v_2 = (3,2), \quad v_3 = (5,2), \quad v_4 = (5,3)$$
$$v_1 = (2,2), \quad v_2 = (3,2), \quad v_3 = (3,3), \quad v_4 = (5,3)$$
$$v_1 = (2,2), \quad v_2 = (2,3), \quad v_3 = (3,3), \quad v_4 = (5,3)$$

So

$$\text{prob}_{\{2,3\},\{2\}}(3,5) = \frac{1}{18}\frac{1}{5}\frac{1}{3}\frac{1}{1} + \frac{1}{18}\frac{1}{5}\frac{1}{3}\frac{1}{2} + \frac{1}{18}\frac{1}{5}\frac{1}{4}\frac{1}{2} = \frac{1}{144} \qquad (3.3)$$

On the other hand, consider

$$\{a_i \mid i \in I\} = \{a_2, a_3\} = \{4, 2\}$$

and

$$\{b_j \mid j \in J\} = \{b_2\} = \{1\}$$

Thus

$$\frac{1}{n} \prod_{i \in I} \frac{1}{a_i} \prod_{j \in J} \frac{1}{b_j} = \frac{1}{18} \frac{1}{4 \cdot 2 \cdot 1} = \frac{1}{144}$$

as in equation (3.3). This is not an accident. In fact, to prove (3.2) (thus finishing the proof of Proposition 3.1.3 and the hook formula), it suffices to show the following.

Lemma 3.1.4 *Let* (α, β), $I = \{i_1, i_2, \ldots\}$ *and* $J = \{j_1, j_2, \ldots\}$ *be given. Then*

$$\text{prob}_{I,J}(\alpha, \beta) = \frac{1}{n} \prod_{i \in I} \frac{1}{a_i} \prod_{j \in J} \frac{1}{b_j}$$

Proof. Suppose first that I or J is empty. For definiteness, take $J = \emptyset$. Then the only possible trial is

$$v_1 = (i_1, \beta), v_2 = (i_2, \beta), \ldots$$

Thus

$$\text{prob}_{I,J}(\alpha, \beta) = \frac{1}{n a_{i_1} a_{i_2} \cdots}$$

as desired.

Now take $I, J \neq \emptyset$. Thus $v_1 = (i_1, j_1)$ and there are exactly two choices for v_2—namely, (i_2, j_1) or (i_1, j_2) (replacing i_2 by α if $|I| = 1$ and similarly for j_2). Let $\bar{I} = I - \{i_1\}$ and $\bar{J} = J - \{j_1\}$. Then, by induction on $|I \cup J|$,

$$
\begin{aligned}
\text{prob}_{I,J}(\alpha, \beta) &= \frac{1}{h_{i_1,j_1} - 1} \left[\text{prob}_{\bar{I},J}(\alpha, \beta) + \text{prob}_{I,\bar{J}}(\alpha, \beta) \right] \\
&= \frac{1}{h_{i_1,j_1} - 1} \left[\frac{1}{n \hat{a}_{i_1} a_{i_2} \cdots b_{j_1} b_{j_2} \cdots} + \frac{1}{n a_{i_1} a_{i_2} \cdots \hat{b}_{j_1} b_{j_2} \cdots} \right] \\
&= \frac{a_{i_1} + b_{j_1}}{h_{i_1,j_1} - 1} \left[\frac{1}{n a_{i_1} a_{i_2} \cdots b_{j_1} b_{j_2} \cdots} \right]
\end{aligned}
$$

where ^ means that the factor is deleted. But

$$h_{i_1,j_1} - 1 = (h_{i_1,\beta} - 1) + (h_{\alpha,j_1} - 1) = a_{i_1} + b_{j_1}$$

so the lemma is proved. ∎

There are many different proofs of the hook formula. Unfortunately, there is no really simple proof that is purely bijective. Franzblau and Zeilberger [F-Z 82] were the first to come up with a (complicated) bijection. Later Zeilberger [Zei 84] gave a bijective version of the Greene-Nijenhuis-Wilf proof, which is still fairly complex. Also, Remmel [Rem 82] has used the Garsia-Milne involution principle [G-M 81] to produce bijections.

3.2   The Determinantal Formula

The determinantal formula for f^λ is a lot older than the hook formula. In fact, the former was known to Frobenius [Fro 00, Fro 03] and Young [You 02]. In the following theorem we set $1/r! = 0$ if $r < 0$.

Theorem 3.2.1 (Determinantal Formula) *If $(\lambda_1, \lambda_2, \ldots, \lambda_l) \vdash n$, then*

$$f^\lambda = n! \left| \frac{1}{(\lambda_i - i + j)!} \right|$$

where the determinant is l by l.

To remember the denominators in the determinant, note that the parts of λ occur along the main diagonal. The other entries in a given row are found by decreasing or increasing the number inside the factorial by 1 for every step taken to the left or right, respectively. Applying this result to $\lambda = (2, 2, 1)$, we get

$$f^{(2,2,1)} = 5! \cdot \begin{vmatrix} 1/2! & 1/3! & 1/4! \\ 1/1! & 1/2! & 1/3! \\ 0 & 1/0! & 1/1! \end{vmatrix} = 5$$

which agrees with our computation in Section 3.1.

Proof (of the determinantal formula). It suffices to show that the determinant equals the denominator of the hook formula. From our definitions $\lambda_i + l = h_{i,1} + i$, so

$$\left| \frac{1}{(\lambda_i - i + j)!} \right| = \left| \frac{1}{(h_{i,1} - l + j)!} \right|$$

Since every row of this determinant is of the form

$$\left[\frac{1}{(h - l + 1)!} \quad \cdots \quad \frac{1}{(h - 2)!} \quad \frac{1}{(h - 1)!} \quad \frac{1}{h!} \right]$$

⌣ we will use only elementary column operations, we will display only the first row in what follows.

$$\left| \frac{1}{(h_{i,1} - l + j)!} \right| = \begin{vmatrix} \frac{1}{(h_{1,1}-l+1)!} & \cdots & \frac{1}{(h_{1,1}-2)!} & \frac{1}{(h_{1,1}-1)!} & \frac{1}{h_{1,1}!} \\ & & \vdots & & \end{vmatrix}$$

$$= \prod_{i=1}^{l} \frac{1}{h_{i,1}!} \cdot \begin{vmatrix} h_{1,1}(h_{1,1}-1)\cdots(h_{1,1}-l+2) & \cdots & h_{1,1}(h_{1,1}-1) & h_{1,1} & 1 \\ & \vdots & & & \end{vmatrix}$$

$$= \prod_{i=1}^{l} \frac{1}{h_{i,1}!} \cdot \begin{vmatrix} h_{1,1}(h_{1,1}-1)\cdots(h_{1,1}-l+2) & \cdots & h_{1,1}(h_{1,1}-1) & h_{1,1}-1 & 1 \\ & \vdots & & & \end{vmatrix}$$

$$= \prod_{i=1}^{l} \frac{1}{h_{i,1}!} \cdot \begin{vmatrix} h_{1,1}(h_{1,1}-1)\cdots(h_{1,1}-l+2) & \cdots & (h_{1,1}-1)(h_{1,1}-2) & h_{1,1}-1 & 1 \\ & \vdots & & & \end{vmatrix}$$

$$= \prod_{i=1}^{l} \frac{1}{h_{i,1}!} \cdot \begin{vmatrix} (h_{1,1}-1)(h_{1,1}-2)\cdots(h_{1,1}-l+1) & \cdots & (h_{1,1}-1)(h_{1,1}-2) & h_{1,1}-1 & 1 \\ & \vdots & & & \end{vmatrix}$$

$$= \prod_{i=1}^{l} \frac{1}{h_{i,1}} \cdot \begin{vmatrix} \frac{1}{(h_{1,1}-l)!} & \cdots & \frac{1}{(h_{1,1}-3)!} & \frac{1}{(h_{1,1}-2)!} & \frac{1}{(h_{1,1}-1)!} \\ & \vdots & & & \end{vmatrix}$$

But, by induction on n, this last determinant is just $1/\prod_{v \in \overline{\lambda}} h_v$, where

$$\begin{aligned} \overline{\lambda} &= (\lambda_1 - 1, \lambda_2 - 1, \ldots, \lambda_l - 1) \\ &= \lambda \text{ with its first column removed} \end{aligned}$$

Note that $\overline{\lambda}$ may no longer have l rows, even though the determinant is still $l \times l$. However induction can still apply since the portion of the determinant corresponding to rows of $\overline{\lambda}$ of length zero is upper triangular with ones on the diagonal. Now $\overline{\lambda}$ contains all the hooklengths $h_{i,j}$ in λ for $i \geq 2$. Thus

$$\prod_{i=1}^{l} \frac{1}{h_{i,1}} \cdot \prod_{v \in \overline{\lambda}} \frac{1}{h_v} = \prod_{v \in \lambda} \frac{1}{h_v}$$

as desired. ∎

Notice that the same argument in reverse can be used to prove the hook formula for f^λ from the determinantal one. This was Frame, Robinson, and Thrall's original method of proof [FRT 54, pages 317–318], except that they started from a slightly different version of the Frobenius-Young formula (see Exercise 3).

3.3 The Robinson-Schensted Algorithm

If we disregard its genesis, the identity

$$\sum_{\lambda \vdash n} (f^\lambda)^2 = n!$$

can be regarded as a purely combinatorial statement. It says that the number of elements in S_n is equal to the number of pairs of standard tableaux of the same shape λ as λ varies over all partitions of n. Thus it should be possible to give a purely combinatorial—i.e., bijective—proof of this formula. The Robinson-Schensted correspondence does exactly that. It was originally discovered by Robinson [Rob 38] and then found independently in quite a different form by Schensted [Sch 61]. It is the latter's version of the algorithm that we present.

The bijection is denoted

$$\pi \stackrel{\text{R--S}}{\longleftrightarrow} (P, Q)$$

where $\pi \in S_n$ and P, Q are standard λ-tableaux, $\lambda \vdash n$. We first exhibit the map that, given a permutation, produces a pair of tableaux.

$\underline{\pi \stackrel{\text{R--S}}{\longrightarrow} (P, Q)}$. Suppose that π is given in two-line notation as

$$\pi = \begin{matrix} 1 & 2 & \cdots & n \\ x_1 & x_2 & \cdots & x_n \end{matrix}$$

We construct a sequence of tableaux

$$(P_0, Q_0) = (\emptyset, \emptyset), \ (P_1, Q_1), \ (P_2, Q_2), \ \ldots, \ (P_n, Q_n) = (P, Q) \qquad (3.4)$$

where $x_1, x_2, \ldots, x_n$ are *inserted* into the P's and $1, 2, \ldots, n$ are *placed* in the Q's so that $\text{sh} \, P_k = \text{sh} \, Q_k$ for all k. The operations of insertion and placement are described as follows.

Suppose P is a partial tableau and let x be an element not in P. To *row insert x into P*, we proceed as follows.

RS1 Set $R :=$ the first row of P.

RS2 **While** x is less than some element of row R, **do**

        begin

    RSa Let y be the smallest element of R greater than x and replace y by x in R (denoted $R \leftarrow x$).

    RSb Set $x := y$ and $R :=$ next row down.

        end

RS3 Now x is greater than every element of R, so place x at the end of row R and **stop**.

To illustrate, suppose $x = 3$ and

$$P = \begin{matrix} 1 \ 2 \ 5 \ 8 \\ 4 \ 7 \\ 6 \\ 9 \end{matrix}$$

To follow the path of the insertion of x into P, we put elements that are displaced (or *bumped*) during the insertion in boldface type.

$$
\begin{array}{llll}
\begin{array}{l} 1\ 2\ 5\ 8 \\ 4\ 7 \\ 6 \\ 9 \end{array} \leftarrow 3 &
\begin{array}{l} 1\ 2\ \mathbf{3}\ 8 \\ 4\ 7 \\ 6 \\ 9 \end{array} \leftarrow 5 &
\begin{array}{l} 1\ 2\ 3\ 8 \\ 4\ \mathbf{5} \\ 6 \\ 9 \end{array} \leftarrow 7 &
\begin{array}{l} 1\ 2\ 3\ 8 \\ 4\ 5 \\ 6\ \mathbf{7} \\ 9 \end{array}
\end{array}
$$

If the result of row inserting x into P yields the tableau P', then write

$$r_x(P) = P'$$

Note that the insertion rules have been carefully written so that P' still has increasing rows and columns.

Placement of an element in a tableaux is even easier than insertion. Suppose that Q is a partial tableau of shape μ and that (i, j) is an outer corner of μ. If k is greater than every element of Q, then to *place k in Q at cell (i, j)*, merely set $Q_{i,j} := k$. The restriction on k guarantees that the new array is still a partial tableau. For example, if we take

$$
Q = \begin{array}{l} 1\ 2\ 5 \\ 4\ 7 \\ 6 \\ 8 \end{array}
$$

then placing $k = 9$ in cell $(i, j) = (2, 3)$ yields

$$
\begin{array}{l} 1\ 2\ 5 \\ 4\ 7\ 9 \\ 6 \\ 8 \end{array}
$$

At last we can describe how to build the sequence (3.4) from the permutation

$$
\pi = \begin{array}{cccc} 1 & 2 & \cdots & n \\ x_1 & x_2 & \cdots & x_n \end{array}
$$

Start with a pair (P_0, Q_0) of empty tableaux. Assuming that (P_{k-1}, Q_{k-1}) has been constructed, define (P_k, Q_k) by

$$
\begin{aligned}
P_k &= r_{x_k}(P_{k-1}) \\
Q_k &= \text{place } k \text{ into } Q_{k-1} \text{ at the cell } (i, j) \text{ where the} \\
&\quad\ \text{insertion terminates}
\end{aligned}
$$

Note that the definition of Q_k ensures that $\operatorname{sh} P_k = \operatorname{sh} Q_k$ for all k. We call $P = P_n$ the *P-tableau*, or *insertion tableau*, of π and write $P = P(\pi)$. Similarly, $Q = Q_n$ is called the *Q-tableau*, or *recording tableau*, and denoted $Q = Q(\pi)$.

Now we consider an example of the complete algorithm. Boldface numbers are used to distinguish the elements of the lower line of π and hence also for the elements of the P_k. Let

$$\pi = \begin{matrix} 1 & 2 & 3 & 4 & 5 & 6 & 7 \\ \mathbf{4} & \mathbf{2} & \mathbf{3} & \mathbf{6} & \mathbf{5} & \mathbf{1} & \mathbf{7} \end{matrix} \tag{3.5}$$

Then the tableaux constructed by the algorithm are:

$$P_k : \quad \begin{matrix} \emptyset, & \mathbf{4}, & \mathbf{2}, & \mathbf{2\ 3}, & \mathbf{2\ 3\ 6}, & \mathbf{2\ 3\ 5}, & \mathbf{1\ 3\ 5}, & \mathbf{1\ 3\ 5\ 7} \\ & & \mathbf{4} & \mathbf{4} & \mathbf{4} & \mathbf{4\ 6} & \mathbf{2\ 6} & \mathbf{2\ 6} \\ & & & & & & \mathbf{4} & \mathbf{4} \end{matrix} \quad = P$$

$$Q_k : \quad \begin{matrix} \emptyset, & 1, & 1, & 1\ 3, & 1\ 3\ 4, & 1\ 3\ 4, & 1\ 3\ 4, & 1\ 3\ 4\ 7 \\ & & 2 & 2 & 2 & 2\ 5 & 2\ 5 & 2\ 5 \\ & & & & & & 6 & 6 \end{matrix} \quad = Q$$

So

$$\begin{matrix} 1 & 2 & 3 & 4 & 5 & 6 & 7 \\ \mathbf{4} & \mathbf{2} & \mathbf{3} & \mathbf{6} & \mathbf{5} & \mathbf{1} & \mathbf{7} \end{matrix} \xrightarrow{\text{R-S}} \left(\begin{matrix} \mathbf{1\ 3\ 5\ 7} \\ \mathbf{2\ 6} \\ \mathbf{4} \end{matrix} \;,\; \begin{matrix} 1\ 3\ 4\ 7 \\ 2\ 5 \\ 6 \end{matrix} \right)$$

The main theorem about this correspondence is as follows.

Theorem 3.3.1 ([Rob 38, Sch 61]) *The map*

$$\pi \xrightarrow{\text{R-S}} (P, Q)$$

just defined is a bijection between elements of S_n and pairs of standard tableau of the same shape $\lambda \vdash n$.

Proof. To show that we have a bijection, it suffices to create an inverse.
$(P, Q) \xrightarrow{\text{S-R}} \pi$. We merely reverse the preceding algorithm step by step. We begin by defining $(P_n, Q_n) = (P, Q)$. Assuming that (P_k, Q_k) has been constructed, we will find x_k (the kth element of π) and (P_{k-1}, Q_{k-1}). To avoid double-subscripting in what follows, we use $P_{i,j}$ to stand for the (i, j) entry of P_k.

Find the cell (i, j) containing k in Q_k. Since this is the largest element in Q_k, $P_{i,j}$ must have been the last element to be displaced in the construction of P_k. We can now use the following procedure to *delete* $P_{i,j}$ from P. For convenience, we assume the existence of an empty zeroth row above the first row of P_k.

SR1 Set $x := P_{i,j}$ and erase $P_{i,j}$.
     Set $R :=$ the $(i-1)^{\text{st}}$ row of P_k.

SR2 **While** R is not the zeroth row of P_k, **do**

          begin

SRa Let y be the largest element of R smaller than x and replace y by x in R.

SRb Set $x := y$ and $R :=$ next row up.

   end

SR3 Now x has been removed from the first row, so set $x_k := x$.

It is easy to see that P_{k-1} is P_k after the deletion process just described is complete and Q_{k-1} is Q_k with the k erased. Continuing in this way, we eventually recover all the elements of π in reverse order. ∎

The Robinson-Schensted algorithm has many surprising and beautiful properties. The rest of this chapter is devoted to discussing some of them.

3.4 Column Insertion

Obviously we can define *column insertion* of x into P by replacing *row* by *column* everywhere in RS1-3. If column insertion of x into P yields P', we write

$$c_x(P) = P'$$

It turns out that the column and row insertion operators commute. Before we can prove this, however, we need a lemma about the insertion path. The reader should be able to supply the details of the proof.

Lemma 3.4.1 *Let P be a partial tableau with $x \notin P$. Suppose that during the insertion $r_x(P) = P'$, the elements $x', x'', x''', \ldots$ are bumped from cells $(1, j'), (2, j''), (3, j'''), \ldots$, respectively. Then*

1. $x < x' < x'' < \cdots$

2. $j' \geq j'' \geq j''' \geq \cdots$

3. $P'_{i,j} \leq P_{i,j}$ *for all i, j* ∎

Proposition 3.4.2 ([Sch 61]) *For any partial tableau P and distinct elements $x, y \notin P$,*

$$c_y r_x(P) = r_x c_y(P)$$

Proof. Let m be the maximum element in $\{x, y\} \cup P$. Then m cannot displace any entry during any of the insertions. The proof breaks into cases depending on where m is located.

Case 1: $y = m$. (The case where $x = m$ is similar.) Represent P schematically as

$$P = \qquad \qquad \qquad \qquad \qquad \qquad \text{(3.6)}$$

Since y is maximal, c_y applied to either P or $r_x(P)$ merely inserts y at the end of the first column. Let $\overline{x}$ be the last element to be bumped during the insertion $r_x(P)$ and suppose it comes to rest in cell u. If u is in the first column, then

$$c_y r_x(P) = \qquad \text{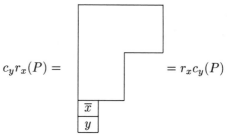} \qquad = r_x c_y(P)$$

If u is in any other column, then both $c_y r_x(P)$ and $r_x c_y(P)$ are equal to

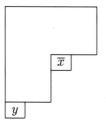

Case 2: $m \in P$. We induct on the number of entries in P. Let $\overline{P}$ be P with the m erased. Then $c_y r_x(\overline{P}) \subset c_y r_x(P)$ and $r_x c_y(\overline{P}) \subset r_x c_y(P)$. But $c_y r_x(\overline{P}) = r_x c_y(\overline{P})$ by induction. Thus $c_y r_x(P)$ and $r_x c_y(P)$ agree everywhere except possibly on the location of m.

To show that m occupies the same position in both tableaux, let $\overline{x}$ be the last element displaced during $r_x(\overline{P})$, say into cell u. Similarly, define $\overline{y}$ and v for the insertion $c_y(\overline{P})$. We now have two subcases, depending on whether u and v are equal or not.

Subcase 2a: $u = v$. Represent $\overline{P}$ as in diagram (3.6). Then $r_x(\overline{P})$ and $c_y(\overline{P})$ are represented, respectively, by

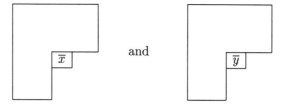

By the previous lemma, the first insertion displaces only elements in columns (weakly) to the right of $u = v$ and the second affects only those in rows (weakly) below. Thus r_x follows the same insertion path when applied to both $\overline{P}$ and $c_y(\overline{P})$ until $\overline{x}$ is bumped from the row above u. A similar statement holds for c_y in $\overline{P}$ and $r_x(\overline{P})$. Thus if $\overline{x} < \overline{y}$ (the case $\overline{x} > \overline{y}$ being similar),

then

$$c_y r_x(\overline{P}) = \quad \boxed{\begin{array}{c} \\ \boxed{\overline{x}} \\ \boxed{\overline{y}} \end{array}} \quad = r_x c_y(\overline{P})$$

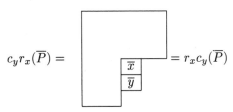

Note that $\overline{x}$ and $\overline{y}$ must end up in the same column.

Now consider what happens to m. If m is not in cell $u = v$, then none of the insertions displace it. So $c_y r_x(P)$ and $r_x c_y(P)$ are both equal to the previous diagram with m added in its cell. If m occupies u, then both $c_y r_x(P)$ and $r_x c_y(P)$ equal the preceding diagram with m added in the column just to the right of u. (In both cases m is bumped there by $\overline{y}$, and in the former it is first bumped one row down by $\overline{x}$.)

Subcase 2b: $u \neq v$. Let $\overline{P}$ have shape $\overline{\lambda}$ and let $c_y r_x(\overline{P}) = r_x c_y(\overline{P})$ have shape λ. Comparing $r_x(\overline{P})$ with $c_y r_x(\overline{P})$ and $c_y(\overline{P})$ with $r_x c_y(\overline{P})$, we see that $\overline{\lambda} \cup \{u\} \subset \lambda$ and $\overline{\lambda} \cup \{v\} \subset \lambda$. Thus

$$\lambda = \overline{\lambda} \cup \{u, v\}$$

Since the insertion paths for $r_x(\overline{P})$ and $c_y(\overline{P})$ may cross, $\overline{x}$ and $\overline{y}$ are not necessarily the entries of cells u and v in $c_y r_x(\overline{P}) = r_x c_y(\overline{P})$. However, one can still verify that u is filled by r_x and v by r_y, whatever the order of the insertions. Consideration of all possible places where m could occur in P now completes the proof. The details are similar to the end of Subcase 2a and are left to the reader. ∎

As an application of this previous proposition, let us consider the *reversal* of π, denoted π^r; i.e., if $\pi = x_1 x_2 \ldots x_n$, then $\pi^r = x_n x_{n-1} \ldots x_1$. The P-tableaux of π and π^r are closely related.

Theorem 3.4.3 ([Sch 61]) *If $P(\pi) = P$, then $P(\pi^r) = P^t$, where t denotes transposition.*

Proof. We have

$$\begin{aligned} P(\pi^r) &= r_{x_1} \cdots r_{x_{n-1}} r_{x_n}(\emptyset) & \text{(definition of } P(\pi^r)) \\ &= r_{x_1} \cdots r_{x_{n-1}} c_{x_n}(\emptyset) & \text{(initial tableau is empty)} \\ &= c_{x_n} r_{x_1} \cdots r_{x_{n-1}}(\emptyset) & \text{(Proposition 3.4.2)} \\ &\ \ \vdots \\ &= c_{x_n} c_{x_{n-1}} \cdots c_{x_1}(\emptyset) & \text{(induction)} \\ &= P^t & \text{(definition of column insertion)} \ \blacksquare \end{aligned}$$

We will characterize the Q-tableau of π^r using Schützenberger's operation of evacuation (Theorem 3.11.4).

If we represent the elements of $\mathcal{S}_n$ as permutation matrices, then the dihedral group of the square (all reflections and rotations bringing a square back onto itself) acts on them. For example, reversal is just reflection in the vertical axis. It would be interesting to see what each of these changes in the permutation does to the output tableaux of the Robinson-Schensted algorithm. This theme motivates us in several of the following sections.

3.5   Increasing and Decreasing Subsequences

One of Schensted's main motivations for constructing his insertion algorithm, as the title of [Sch 61] suggests, was to study increasing and decreasing subsequences of a given sequence $\pi \in \mathcal{S}_n$. It turns out that these are intimately connected with the tableau $P(\pi)$.

Definition 3.5.1 Given $\pi = x_1 x_2 \ldots x_n \in \mathcal{S}_n$. Then an *increasing* (respectively, *decreasing*) *subsequence of* π is

$$x_{i_1} < x_{i_2} < \cdots < x_{i_k}$$

(respectively,

$$x_{i_1} > x_{i_2} > \cdots > x_{i_k})$$

where $i_1 < i_2 < \cdots < i_k$. The integer k is the *length* of the subsequence. ∎

By way of illustration, take the permutation

$$\pi = 4\ 2\ 3\ 6\ 5\ 1\ 7 \tag{3.7}$$

Then an increasing subsequence of π is 2 3 5 7 of length 4, and a decreasing subsequence is 4 3 1 of length 3. In fact, it is easy to check that these are the longest increasing and decreasing subsequences of π. On page 100 we found the P-tableau of π to be

$$P(\pi) = \begin{matrix} 1\ 3\ 5\ 7 \\ 2\ 6 \\ 4 \end{matrix} \tag{3.8}$$

Note that the length of the first row of $P(\pi)$ is 4, whereas the length of the first column is 3. This is not a coincidence, as the next theorem shows.

Theorem 3.5.2 ([Sch 61]) *Consider* $\pi \in \mathcal{S}_n$. *The length of the longest increasing subsequence of* π *is the length of the first row of* $P(\pi)$. *The length of the longest decreasing subsequence of* π *is the length of the first column of* $P(\pi)$.

Proof. Since reversing a permutation turns decreasing sequences into increasing ones, the second half of the theorem follows from the first and Theorem 3.4.3. To prove the first half, we actually demonstrate a stronger result. In what follows, P_{k-1} is the the tableau formed after $k - 1$ insertions of the Robinson-Schensted algorithm.

Lemma 3.5.3 *If* $\pi = x_1 x_2 \ldots x_n$ *and* x_k *enters* P_{k-1} *in column* j, *then the longest increasing subsequence of* π *ending in* x_k *has length* j.

Proof. We induct on k. The result is trivial for $k = 1$, so suppose it holds for all values up to $k - 1$.

First we need to show the existence of an increasing subsequence of length j ending in x_k. Let y be the element of P_{k-1} in cell $(1, j - 1)$. Then we have $y < x_k$, since x_k enters in column j. Also, by induction, there is an increasing subsequence σ of length $j-1$ ending in y. Thus σx_k is the desired subsequence.

Now we must prove that there cannot be a longer increasing subsequence ending in x_k. Suppose that such a subsequence exists and let x_i be the element preceding x_k in that subsequence. Then, by induction, when x_i is inserted, it enters in some column (weakly) to the right of column j. Thus the element y in cell $(1, j)$ of P_i satisfies

$$y \leq x_i < x_k$$

But by part 3 of Lemma 3.4.1, the entries in a given position of a tableau never increase with subsequent insertions. Thus the element in cell $(1, j)$ of P_{k-1} is smaller than x_k, contradicting the fact that x_k displaces it. This finishes the proof of the lemma and hence of Theorem 3.5.2. ∎

Note that the first row of $P(\pi)$ need not be an increasing subsequence of π even though it has the right length; compare (3.7) and (3.8). However, an increasing subsequence of π of maximum length can be reconstructed from the Robinson-Schensted algorithm.

It turns out that an interpretation can be given to the lengths of the other rows and columns of $P(\pi)$, and we will do this in Section 3.7. First, however, we must develop an appropriate tool for the proof.

3.6 The Knuth Relations

Suppose we wish to prove a theorem about the elements of a set that is divided into equivalence classes. Then a common way to accomplish this is to show that

1. the theorem holds for a particular element of each equivalence class, and

2. if the theorem holds for one element of an equivalence class, then it holds for all elements of the class.

The set $\mathcal{S}_n$ has the following useful equivalence relation on it.

Definition 3.6.1 Two permutations $\pi, \sigma \in \mathcal{S}_n$ are said to be *P-equivalent*, written $\pi \stackrel{P}{\cong} \sigma$, if $P(\pi) = P(\sigma)$. ∎

For example, the equivalence classes in S_3 are

$$\{1\ 2\ 3\} \quad \{2\ 1\ 3,\ 2\ 3\ 1\} \quad \{1\ 3\ 2, 3\ 1\ 2\} \quad \{3\ 2\ 1\}$$

corresponding to the P-tableaux

$$
1\ 2\ 3 \qquad \begin{matrix} 1\ 3 \\ 2 \end{matrix} \qquad \begin{matrix} 1\ 2 \\ 3 \end{matrix} \qquad \begin{matrix} 1 \\ 2 \\ 3 \end{matrix}
$$

respectively.

We can prove a strengthening of Schensted's theorem 3.5.2 using this equivalence relation and the proof technique just outlined. However, we also need an alternative description of P-equivalence. This is given by the Knuth relations.

Definition 3.6.2 Suppose $x < y < z$. Then $\pi, \sigma \in S_n$ *differ by a Knuth relation of the first kind*, written $\pi \overset{1}{\cong} \sigma$, if

    1. $\pi = x_1 \ldots yxz \ldots x_n$ and $\sigma = x_1 \ldots yzx \ldots x_n$ or vice versa

They *differ by a Knuth relation of the second kind*, written $\pi \overset{2}{\cong} \sigma$, if

    2. $\pi = x_1 \ldots xzy \ldots x_n$ and $\sigma = x_1 \ldots zxy \ldots x_n$ or vice versa

The two permutations are *Knuth equivalent*, written $\pi \overset{K}{\cong} \sigma$, if there is a sequence of permutations such that

$$
\pi = \pi_1 \overset{i}{\cong} \pi_2 \overset{j}{\cong} \cdots \overset{l}{\cong} \pi_k = \sigma
$$

where $i, j, \ldots, l \in \{1, 2\}$. ∎

Returning to S_3 we see that the only nontrivial Knuth relations are

$$
2\ 1\ 3 \overset{1}{\cong} 2\ 3\ 1 \quad \text{and} \quad 1\ 3\ 2 \overset{2}{\cong} 3\ 1\ 2
$$

Thus the Knuth equivalence classes and P-equivalence classes coincide. This always happens.

Theorem 3.6.3 ([Knu 70]) *If $\pi, \sigma \in S_n$, then*

$$
\pi \overset{K}{\cong} \sigma \iff \pi \overset{P}{\cong} \sigma
$$

Proof. $\implies$ It suffices to prove that $\pi \overset{P}{\cong} \sigma$ whenever π and σ differ by a Knuth relation. In fact, the result for $\overset{2}{\cong}$ follows from the one for $\overset{1}{\cong}$ because

$$
\begin{aligned}
\pi \overset{2}{\cong} \sigma \implies & \pi^r \overset{1}{\cong} \sigma^r && \text{(by definitions)} \\
\implies & P(\pi^r) = P(\sigma^r) && \text{(proved later)} \\
\implies & P(\pi)^t = P(\sigma)^t && \text{(Theorem 3.4.3)} \\
\implies & P(\pi) = P(\sigma)
\end{aligned}
$$

Now assume $\pi \stackrel{1}{\cong} \sigma$. Keeping the notation of Definition 3.6.2, let P be the tableau obtained by inserting the elements before y (which are the same in both π and σ). Thus it suffices to prove

$$r_z r_x r_y(P) = r_x r_z r_y(P)$$

Intuitively, what we will show is that the insertion path of y creates a "barrier," so that the paths for x and z lie to the left and right of this line, respectively, no matter in which order they are inserted. Since these two paths do not intersect, $r_z r_x$ and $r_x r_z$ have the same effect.

We induct on the number of rows of P. If P has no rows—i.e., $P = \emptyset$—then it is easy to verify that both sequences of insertions yield the tableau

$$x \quad z$$
$$y$$

Now let P have $r > 0$ rows and consider $\overline{P} = r_y(P)$. Suppose y enters P in column k, displacing element y'. The crucial facts about $\overline{P}$ are

1. $\overline{P}_{1,j} \leq y$ for all $j \leq k$, and

2. $\overline{P}_{1,l} > y'$ for all $l > k$.

If the insertion of x is performed next, then, since $x < y$, x must enter in some column j with $j \leq k$. Furthermore, if x' is the element displaced, then we must have $x' < y'$ because of our first crucial fact. Applying r_z to $r_x r_y(P)$ forces $z > y$ to come in at column l, where $l > k$ for the same reason. Also the element z' displaced satisfies $z' > y'$ by crucial fact 2.

Now consider $r_x r_z r_y(P)$. Since the crucial facts continue to hold, z and x must enter in columns strictly to the right and weakly to the left of column k, respectively. Because these two sets of columns are disjoint, the entrance of one element does not effect the entrance of the other. Thus the first rows of $r_z r_x r_y(P)$ and $r_x r_z r_y(P)$ are equal. In addition, the elements displaced in both cases are x', y', and z', satisfying $x' < y' < z'$. Thus if P' denotes the bottom $r - 1$ rows of P, then the rest of our two tableaux can be found by computing $r_{z'} r_{x'} r_{y'}(P')$ and $r_{x'} r_{z'} r_{y'}(P')$. These last two arrays are equal by induction, so we are done with this half of the proof.

Before continuing with the other implication, we need to introduce some more concepts.

Definition 3.6.4 If P is a tableau, then the *row word of P* is the permutation

$$\pi_P = R_l R_{l-1} \ldots R_1$$

where $R_1, \ldots, R_l$ are the rows of P. ∎

For example, if

$$P = \begin{array}{l} 1\ 3\ 5\ 7 \\ 2\ 6 \\ 4 \end{array}$$

then
$$\pi_P = 4\ 2\ 6\ 1\ 3\ 5\ 7$$

The following lemma is easy to verify directly from the definitions.

Lemma 3.6.5 *If P is a standard tableau, then*

$$\pi_P \xrightarrow{\text{R-S}} (P, \cdot)\ \blacksquare$$

The reader may have noticed that most of the definitions and theorems above involving standard tableaux and permutations make equally good sense when applied to partial tableaux and *partial permutations*, which are bijections $\pi : K \to L$ between two sets of positive integers. If $K = \{k_1 < k_2 < \cdots < k_m\}$, then we can write π in two-line form as

$$\pi = \begin{matrix} k_1 & k_2 & \cdots & k_m \\ l_1 & l_2 & \cdots & l_m \end{matrix}$$

where $l_i = \pi(k_i)$. Insertion of the k_i and placement of the l_i set up a bijection $\pi \xrightarrow{\text{R-S}} (P, Q)$ with partial tableaux. Henceforth, we assume the more general case whenever it suits us to do so. Now back to the proof of Theorem 3.6.3.

$\Longleftarrow$ By transitivity of equivalence relations, it suffices to show that

$$\pi \stackrel{K}{\cong} \pi_P$$

whenever $P(\pi) = P$.

We induct on n, the number of elements in π. Let x be the last element of π so that $\pi = \pi'x$, where π' is a sequence of $n-1$ integers. Let $P' = P(\pi')$. Then, by induction,

$$\pi' \stackrel{K}{\cong} \pi_{P'}$$

So it suffices to prove that

$$\pi_{P'}x \stackrel{K}{\cong} \pi_P$$

In fact, we will show that the Knuth relations used to transform $\pi_{P'}x$ into π_P essentially simulate the insertion of x into the tableau P' (which of course yields the tableau P).

Let the rows of P' be $R_1, \ldots, R_l$, where $R_1 = p_1p_2\ldots p_k$. If x enters P' in column j, then

$$p_1 < \cdots < p_{j-1} < x < p_j < \cdots < p_k$$

Thus

$$\begin{aligned} \pi_{P'}x &= R_l \ldots R_2 p_1 \ldots p_{j-1}p_j \ldots p_{k-1}p_k x \\ &\stackrel{1}{\cong} R_l \ldots R_2 p_1 \ldots p_{j-1}p_j \ldots p_{k-1}xp_k \\ &\stackrel{1}{\cong} R_l \ldots R_2 p_1 \ldots p_{j-1}p_j \ldots xp_{k-1}p_k \end{aligned}$$

$$\vdots$$
$$\stackrel{1}{\cong} \quad R_l \ldots R_2 p_1 \ldots p_{j-1} p_j x p_{j+1} \ldots p_k$$
$$\stackrel{2}{\cong} \quad R_l \ldots R_2 p_1 \ldots p_j p_{j-1} x p_{j+1} \ldots p_k$$
$$\vdots$$
$$\stackrel{2}{\cong} \quad R_l \ldots R_2 p_j p_1 \ldots p_{j-1} x p_{j+1} \ldots p_k$$

Now the tail of our permutation is exactly the first row of $P = r_x(P')$. Also, the element bumped from the first row of P'—namely p_j—is poised at the end of the sequence corresponding to R_2. Thus we can model its insertion into the second row just as we did with x and R_1. Continuing in this manner, we eventually obtain the row word of P, completing the proof. ∎

3.7 Subsequences Again

Curtis Greene [Gre 74] gave a nice generalization of Schensted's result on increasing and decreasing sequences (Theorem 3.5.2). It involves unions of such sequences.

Definition 3.7.1 Let π be a sequence. A subsequence σ of π is *k-increasing* if, as a set, it can be written as

$$\sigma = \sigma_1 \uplus \sigma_2 \uplus \cdots \uplus \sigma_k$$

where the σ_i are increasing subsequences of π. If the σ_i are all decreasing, then we say that σ is *k-decreasing*. Let

$$i_k(\pi) = \text{the length of } \pi\text{'s longest } k\text{-increasing subsequence}$$

and

$$d_k(\pi) = \text{the length of } \pi\text{'s longest } k\text{-decreasing subsequence} \blacksquare$$

Notice that the case $k = 1$ corresponds to sequences that are merely increasing or decreasing. Using our canonical example permutation (3.7),

$$\pi = 4\,2\,3\,6\,5\,1\,7$$

we see that longest 1-, 2- and 3-increasing subsequences are given by

$$2\,3\,5\,7 \quad \text{and}$$
$$4\,2\,3\,6\,5\,7 \;=\; 2\,3\,5\,7 \uplus 4\,6 \qquad \text{and}$$
$$4\,2\,3\,6\,5\,1\,7 \;=\; 2\,3\,5\,7 \uplus 4\,6 \uplus 1$$

respectively. Thus

$$i_1(\pi) = 4, \qquad i_2(\pi) = 6, \qquad i_3(\pi) = 7 \tag{3.9}$$

Recall that $P(\pi)$ in (3.8) had shape $\lambda = (4, 2, 1)$, so that

$$\lambda_1 = 4, \qquad \lambda_1 + \lambda_2 = 6, \qquad \lambda_1 + \lambda_2 + \lambda_3 = 7 \qquad (3.10)$$

Comparing the values in (3.9) and (3.10), the reader should see a pattern.

To talk about the situation with k-decreasing subsequences conveniently, we need some notation concerning the columns of λ.

Definition 3.7.2 Let λ be a Ferrers diagram; then the *conjugate of* λ is the partition

$$\lambda' = (\lambda'_1, \lambda'_2, \ldots, \lambda'_m)$$

where λ'_i is the length of the ith column of λ. Otherwise put, λ' is just the transpose of the diagram of λ. For this reason it is sometimes denoted λ^t. ∎

For example, if

$$\lambda = (4, 2, 1) = \quad \begin{matrix} \bullet & \bullet & \bullet & \bullet \\ \bullet & \bullet & & \\ \bullet & & & \end{matrix}$$

then

$$\lambda' = \lambda^t = \quad \begin{matrix} \bullet & \bullet & \bullet \\ \bullet & \bullet & \\ \bullet & & \\ \bullet & & \end{matrix} \quad = (3, 2, 1, 1)$$

Greene's theorem is as follows.

Theorem 3.7.3 ([Gre 74]) *Given* $\pi \in S_n$, *let* $\operatorname{sh} P(\pi) = (\lambda_1, \lambda_2, \ldots, \lambda_l)$ *with conjugate* $(\lambda'_1, \lambda'_2, \ldots, \lambda'_m)$. *Then for any* k,

$$\begin{aligned} i_k(\pi) &= \lambda_1 + \lambda_2 + \cdots + \lambda_k \\ d_k(\pi) &= \lambda'_1 + \lambda'_2 + \cdots + \lambda'_k \end{aligned}$$

Proof. By Theorem 3.4.3, we need to prove the statement only for k-increasing subsequences.

We use the proof technique detailed at the beginning of Section 3.6 and the equivalence relation is the one studied there. Given an equivalence class corresponding to a tableau P, use the permutation π_P as the special representative of that class. So we must first prove the result for the row word.

By construction, the first k rows of P form a k-increasing subsequence of π_P, so $i_k(\pi_P) \geq \lambda_1 + \lambda_2 + \cdots + \lambda_k$. To show the reverse inequality holds, note that any k-increasing subsequence can intersect a given decreasing subsequence in at most k elements. Since the columns of P partition π_P into decreasing subsequences,

$$i_k(\pi_P) \leq \sum_{i=1}^{m} \min(\lambda'_i, k) = \sum_{i=1}^{k} \lambda_i$$

Now we must show that the theorem holds for all permutations in the equivalence class of P. Given π in this class, Theorem 3.6.3 guarantees the existence of permutations $\pi_1, \pi_2, \ldots, \pi_j$ such that

$$\pi = \pi_1 \overset{K}{\cong} \pi_2 \overset{K}{\cong} \cdots \overset{K}{\cong} \pi_j = \pi_P$$

where each π_{i+1} differs from π_i by a Knuth relation. Since all the π_i have the same P-tableau, we need to prove only that they all have the same value for i_k. Thus it suffices to show that if $\pi \overset{i}{\cong} \sigma$ for $i = 1, 2$, then $i_k(\pi) = i_k(\sigma)$. We will do the case of $\overset{1}{\cong}$, leaving $\overset{2}{\cong}$ as an exercise for the reader.

Suppose $\pi = x_1 \ldots yxz \ldots x_n$ and $\sigma = x_1 \ldots yzx \ldots x_n$. To prove that $i_k(\pi) \le i_k(\sigma)$, we need to show only that any k-increasing subsequence of π has a corresponding k-increasing subsequence in σ of the same length. Let

$$\pi' = \pi_1 \uplus \pi_2 \uplus \cdots \uplus \pi_k$$

be the subsequence of π. If x and z are not in the same π_i for any i, then the π_i are also increasing subsequences of σ and we are done.

Now suppose that x and z are both in π_1 (the choice of which of the component subsequences is immaterial). If $y \notin \pi'$, then let

$$\sigma_1 = \pi_1 \qquad \text{with } x \text{ replaced by } y$$

Since $x < y < z$, σ_1 is still increasing and

$$\sigma' = \sigma_1 \uplus \pi_2 \uplus \cdots \uplus \pi_k$$

is a subsequence of σ of the right length. Finally, suppose $y \in \pi'$—say $y \in \pi_2$ (note that we can not have $y \in \pi_1$). Let

$$\begin{array}{rcl} \pi_1' & = & \text{the subsequence of } \pi_1 \text{ up to and including } x \\ \pi_1'' & = & \text{the subsequence consisting of the rest of } \pi_1 \\ \pi_2' & = & \text{the subsequence of } \pi_2 \text{ up to and including } y \\ \pi_2'' & = & \text{the subsequence consisting of the rest of } \pi_2 \end{array}$$

Note that $\pi_i = \pi_i' \pi_i''$ for $i = 1, 2$. Construct

$$\sigma_1 = \pi_1' \pi_2'' \quad \text{and} \quad \sigma_2 = \pi_2' \pi_1''$$

which are increasing because $x < y < \min \pi_2''$ and $y < z$, respectively. Also,

$$\sigma' = \sigma_1 \uplus \sigma_2 \uplus \pi_3 \uplus \cdots \uplus \pi_k$$

is a subsequence of σ because x and z are no longer in the same component subsequence. Since its length is correct, we have proved the desired inequality.

To show that $i_k(\pi) \ge i_k(\sigma)$ is even easier. Since x and z are out of order in σ, they can never be in the same component of a k-increasing subsequence.

Thus we are reduced to the first case, and this finishes the proof of the theorem. ∎

The fact that $i_k(\pi) = \lambda_1 + \lambda_2 + \cdots + \lambda_k$ does not imply that we can find a k-increasing subsequence of maximum length

$$\pi' = \pi_1 \uplus \pi_2 \uplus \cdots \uplus \pi_k$$

such that the length of π_i is λ_i for all i. As an example [Gre 74], consider

$$\pi = 2\ 4\ 7\ 9\ 5\ 1\ 3\ 6\ 8$$

Then

$$P(\pi) = \begin{array}{ccccc} 1 & 3 & 4 & 6 & 8 \\ 2 & 4 & 9 & & \\ 7 & & & & \end{array}$$

so $i_2(\pi) = 5 + 3 = 8$. There is only one 2-increasing subsequence of π having length 8—namely,

$$\pi' = 2\ 4\ 7\ 9\ \uplus\ 1\ 3\ 6\ 8$$

The reader can check that it is impossible to represent π' as the disjoint union of two increasing subsequences of lengths 5 and 3.

3.8   Viennot's Geometric Construction

We now return to our study (begun at the end of Section 3.4) of the effects that various changes in a permutation π have on the pair $(P(\pi), Q(\pi))$. We prove a remarkable theorem of Schützenberger [Scü 63] stating that taking the inverse of the permutation merely interchanges the two tableaux; i.e., if $\pi \xrightarrow{\text{R-S}} (P, Q)$, then $\pi^{-1} \xrightarrow{\text{R-S}} (Q, P)$. Our primary tool is a beautiful geometric description of the Robinson-Schensted correspondence due to Viennot [Vie 76].

Consider the first quadrant of the Cartesian plane. Given a permutation $\pi = x_1 x_2 \ldots x_n$, represent x_i by a box with coordinates (i, x_i) (compare this to the permutation matrix of π). Using our running example permutation from equation (3.7),

$$\pi = 4\ 2\ 3\ 6\ 5\ 1\ 7$$

we obtain the following figure.

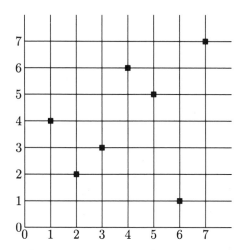

Now imagine a light shining from the origin so that each box casts a shadow with boundaries parallel to the coordinate axes. For example, the shadow cast by the box at (4,6) looks like this:

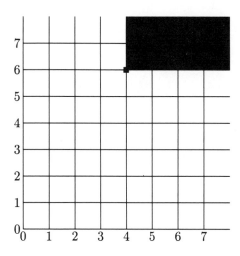

Similar figures result for the other boxes. Consider those points of the permutation that are in the shadow of no other point, in our case (1,4), (2,2), and (6,1). The *first shadow line*, L_1, is the boundary of the combined shadows of these boxes. In the next figure, the appropriate line has been thickened.

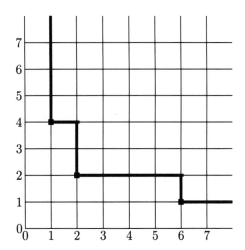

Note that this is a broken line consisting of line segments and exactly one horizontal and one vertical ray.

To form the *second shadow line*, L_2, one removes the boxes on the first shadow line and repeats this procedure.

Definition 3.8.1 Given a permutation displayed in the plane, we form its *shadow lines* $L_1, L_2, \ldots$ as follows. Assuming that $L_1, \ldots, L_{i-1}$ have been constructed, remove all boxes on these lines. Then L_i is the boundary of the shadow of the remaining boxes. The *x-coordinate of L_i* is

$$x_{L_i} = \text{the } x\text{-coordinate of } L_i\text{'s vertical ray}$$

and the *y-coordinate* is

$$y_{L_i} = \text{the } y\text{-coordinate of } L_i\text{'s horizontal ray}$$

The shadow lines make up the *shadow diagram of π*. ∎

In our example there are four shadow lines, and their x- and y-coordinates are shown above and to the left of the following shadow diagram, respectively.

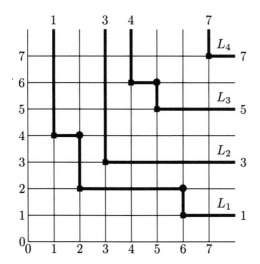

Compare the coordinates of our shadow lines with the first rows of the tableaux

$$P(\pi) = \begin{matrix} 1\ 3\ 5\ 7 \\ 2\ 6 \\ 4 \end{matrix} \qquad \text{and} \qquad Q(\pi) = \begin{matrix} 1\ 3\ 4\ 7 \\ 2\ 5 \\ 6 \end{matrix}$$

computed on page 100. It seems as if

$$P_{1,j} = y_{L_j} \text{ and } Q_{1,j} = x_{L_j} \tag{3.11}$$

for all j. In fact, even more is true. The boxes on line L_j are precisely those elements passing through the $(1, j)$ cell during the construction of P, as the next result shows.

Lemma 3.8.2 *Let the shadow diagram of $\pi = x_1 x_2 \ldots x_n$ be constructed as before. Suppose the vertical line $x = k$ intersects i of the shadow lines. Let y_j be the y-coordinate of the lowest point of the intersection with L_j. Then the first row of the $P_k = P(x_1 \ldots x_k)$ is*

$$R_1 = y_1\ y_2\ \ldots y_i \tag{3.12}$$

Proof. Induct on k, the lemma being trivial for $k = 0$. Assume the result holds for the line $x = k$ and consider $x = k + 1$. There are two cases.
    If

$$x_{k+1} > y_i \tag{3.13}$$

then the box $(k + 1, x_{k+1})$ starts a new shadow line. So none of the values $y_1, \ldots, y_i$ change and we obtain a new intersection,

$$y_{i+1} = x_{k+1}$$

But by (3.12) and (3.13), the $(k+1)^{\text{st}}$ intersection merely causes x_{k+1} to sit at the end of the first row without displacing any other element. Thus the lemma continues to be true.

If, on the other hand,

$$y_1 < \cdots < y_{j-1} < x_{k+1} < y_j < \cdots < y_i \tag{3.14}$$

then $(k+1, x_{k+1})$ is added to line L_j. Thus the lowest coordinate on L_j becomes

$$y_j' = x_{k+1}$$

and all other y-values stay the same. Furthermore, equations (3.12) and (3.14) ensure that the first row of P_{k+1} is

$$y_1 \ \cdots y_{j-1} \ y_j' \ \cdots y_i$$

This is precisely what is predicted by the shadow diagram. ∎

It follows from the proof of the previous lemma that the shadow diagram of π can be read left to right like a time-line recording the construction of $P(\pi)$. At the kth stage, the line $x = k$ intersects one shadow line in a ray or line segment and all the rest in single points. In terms of the first row of P_k: a ray corresponds to placing an element at the end, a line segment corresponds to displacing an element, and the points correspond to elements that are unchanged.

We can now prove that equation (3.11) always holds.

Corollary 3.8.3 ([Vie 76]) *If the permutation π has Robinson-Schensted tableaux (P, Q) and shadow lines L_j, then, for all j,*

$$P_{1,j} = y_{L_j} \quad and \quad Q_{1,j} = x_{L_j}$$

Proof. The statement for P is just the case $k = n$ of Lemma 3.8.2.

As for Q, the entry k is added to Q in cell $(1, j)$ when x_k is greater than every element of the first row of P_{k-1}. But the previous lemma's proof shows that this happens precisely when the line $x = k$ intersects shadow line L_j in a vertical ray. In other words, $y_{L_j} = k = Q_{1,j}$ as desired. ∎

How do we recover the rest of the the P- and Q-tableaux from the shadow diagram of π? Consider the northeast corners of the shadow lines. These are marked with a dot in the diagram on page 114. If such a corner has coordinates (k, x'), then, by the proof of Lemma 3.8.2, x' must be displaced from the first row of P_{k-1} by the insertion of x_k. So the dots correspond to the elements inserted into the second row during the construction of P. Thus we can get the rest of the two tableaux by iterating the shadow diagram construction. In our example, the second and third rows come from the thickened and dashed lines, respectively, of the following diagram.

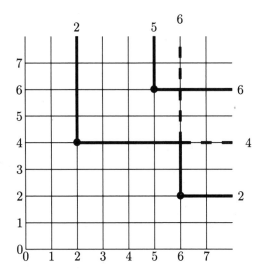

Formally, we have the following definition.

Definition 3.8.4 The ith *skeleton of* $\pi \in S_n$, $\pi^{(i)}$, is defined inductively by $\pi^{(1)} = \pi$ and

$$\pi^{(i)} = \begin{array}{cccc} k_1 & k_2 & \cdots & k_m \\ l_1 & l_2 & \cdots & l_m \end{array}$$

where $(k_1, l_1), \ldots, (k_m, l_m)$ are the northeast corners of the shadow diagram of $\pi^{(i-1)}$ listed in lexicographic order. The shadow lines for $\pi^{(i)}$ are denoted $L_j^{(i)}$. ∎

The next theorem should be clear, given Corollary 3.8.3 and the discussion surrounding Lemma 3.8.2.

Theorem 3.8.5 ([Vie 76]) *Suppose* $\pi \xrightarrow{\text{R-S}} (P, Q)$. *Then* $\pi^{(i)}$ *is a partial permutation such that*

$$\pi^{(i)} \xrightarrow{\text{R-S}} (P^{(i)}, Q^{(i)})$$

where $P^{(i)}$ *(respectively,* $Q^{(i)}$*) consists of the rows* i *and below of* P *(respectively,* Q*). Furthermore,*

$$P_{i,j} = y_{L_j^{(i)}} \quad and \quad Q_{i,j} = x_{L_j^{(i)}}$$

for all i, j. ∎

It is now trivial to demonstrate Schützenberger's theorem

Theorem 3.8.6 ([Scü 63]) *If* $\pi \in S_n$, *then*

$$P(\pi^{-1}) = Q(\pi) \quad and \quad Q(\pi^{-1}) = P(\pi)$$

Proof. Taking the inverse of a permutation corresponds to reflecting the shadow diagram in the line $y = x$. The theorem now follows from Theorem 3.8.5. ∎

As an application of Theorem 3.8.6 we can find those transpositions that, when applied to $\pi \in S_n$, leave $Q(\pi)$ invariant. Dual to our definition of P-equivalence is

Definition 3.8.7 Two permutations $\pi, \sigma \in S_n$ are said to be *Q-equivalent*, written $\pi \overset{Q}{\cong} \sigma$, if $Q(\pi) = Q(\sigma)$. ∎

For example,

$$Q(2\ 1\ 3) = Q(3\ 1\ 2) = \begin{smallmatrix} 1 & 3 \\ 2 & \end{smallmatrix} \quad \text{and} \quad Q(1\ 3\ 2) = Q(2\ 3\ 1) = \begin{smallmatrix} 1 & 2 \\ 3 & \end{smallmatrix}$$

so

$$2\ 1\ 3 \overset{Q}{\cong} 3\ 1\ 2 \text{ and } 1\ 3\ 2 \overset{Q}{\cong} 2\ 3\ 1 \tag{3.15}$$

We also have a dual notion for the Knuth relations.

Definition 3.8.8 Permutations $\pi, \sigma \in S_n$ *differ by a dual Knuth relation of the first kind*, written $\pi \overset{1^*}{\cong} \sigma$, if for some k

1. $\pi = \ldots k+1 \ldots k \ldots k+2 \ldots$ and $\sigma = \ldots k+2 \ldots k \ldots k+1 \ldots$ or vice versa

They *differ by a dual Knuth relation of the second kind*, written $\pi \overset{2^*}{\cong} \sigma$, if for some k

2. $\pi = \ldots k \ldots k+2 \ldots k+1 \ldots$ and $\sigma = \ldots k+1 \ldots k+2 \ldots k \ldots$ or vice versa

The two permutations are *dual Knuth equivalent*, written $\pi \overset{K^*}{\cong} \sigma$, if there is a sequence of permutations such that

$$\pi = \pi_1 \overset{i^*}{\cong} \pi_2 \overset{j^*}{\cong} \cdots \overset{l^*}{\cong} \pi_k = \sigma$$

where $i, j, \ldots, l \in \{1, 2\}$. ∎

Note that the only two nontrivial dual Knuth relations in S_3 are

$$2\ 1\ 3 \overset{1^*}{\cong} 3\ 1\ 2 \quad \text{and} \quad 1\ 3\ 2 \overset{2^*}{\cong} 2\ 3\ 1$$

These correspond exactly to (3.15).

The following lemma is obvious from the definitions. In fact, the definition of the dual Knuth relations was concocted precisely so that this result should hold.

Lemma 3.8.9 *If $\pi, \sigma \in \mathcal{S}_n$, then*

$$\pi \overset{K}{\cong} \sigma \iff \pi^{-1} \overset{K^*}{\cong} \sigma^{-1} \ \blacksquare$$

Now it is an easy matter to derive the dual version of Knuth's theorem about P-equivalence (Theorem 3.6.3).

Theorem 3.8.10 *If $\pi, \sigma \in \mathcal{S}_n$, then*

$$\pi \overset{K^*}{\cong} \sigma \iff \pi \overset{Q}{\cong} \sigma$$

Proof. We have the following string of equivalences

$$
\begin{aligned}
\pi \overset{K^*}{\cong} \sigma \quad &\iff \quad \pi^{-1} \overset{K}{\cong} \sigma^{-1} && \text{(Lemma 3.8.9)} \\
&\iff \quad P(\pi^{-1}) = P(\sigma^{-1}) && \text{(Theorem 3.6.3)} \\
&\iff \quad Q(\pi) = Q(\sigma) && \text{(Theorem 3.8.6)} \ \blacksquare
\end{aligned}
$$

3.9 Schützenberger's Jeu de Taquin

The jeu de taquin (or "teasing game") of Schützenberger [Scü 76] is a powerful tool. It can be used to give alternative descriptions of both the P- and Q-tableaux of the Robinson-Schensted algorithm (Theorems 3.9.7 and 3.11.4) as well as the ordinary and dual Knuth relations (Theorems 3.9.8 and 3.10.8).

To get the full-strength version of these concepts, we must generalize to skew tableaux.

Definition 3.9.1 *If $\mu \subseteq \lambda$ as Ferrers diagrams, then the corresponding* skew *diagram, or* skew shape, *is the set of cells*

$$\lambda/\mu = \{c \mid c \in \lambda \text{ and } c \notin \mu\}$$

A skew diagram is normal *if $\mu = \emptyset$.* ∎

If $\lambda = (3, 3, 2, 1)$ and $\mu = (2, 1, 1)$, then we have the skew diagram

$$\lambda/\mu = \qquad$$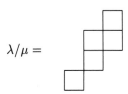

Of course, normal shapes are the left-justified ones we have been considering all along.

The definitions of skew tableau, standard skew tableau, and so on, are all as expected. In particular, the definition of the row word of a tableau still makes sense in this setting. Thus we can say that two skew partial tableaux P, Q are *Knuth equivalent*, written $P \overset{K}{\cong} Q$, if

$$\pi_P \overset{K}{\cong} \pi_Q$$

Similar definitions hold for the other equivalence relations that we have introduced. Note that if $\pi = x_1 x_2 \ldots x_n$, then we can make π into a skew tableau by putting x_i in the cell $(n - i + 1, i)$ for all i. This object is called the *antidiagonal strip tableau associated with π* and is also denoted π. For example, if $\pi = 3142$ (a good approximation, albeit without the decimal point), then

$$\pi = \begin{array}{cccc} & & & \boxed{2} \\ & & \boxed{4} & \\ & \boxed{1} & & \\ \boxed{3} & & & \end{array}$$

So $\pi \overset{K}{\cong} \sigma$ as permutations if and only if $\pi \overset{K}{\cong} \sigma$ as tableaux.

We now come to the definition of a *jeu de taquin slide*, which is essential to all that follows.

Definition 3.9.2 Given a partial tableau P of shape λ/μ, we perform a *forward slide on P into a cell c* as follows.

    F1 Pick c to be an outer corner of λ.

    F2 **While** c is not an outer corner of μ **do**

        Fa If $c = (i, j)$, then let c' be the cell of $\max\{P_{i-1,j}, P_{i,j-1}\}$.
        Fb Slide $P_{c'}$ into cell c and let $c := c'$.

If only one of $P_{i-1,j}, P_{i,j-1}$ exist in step Fa, then the maximum is taken to be that single value. We denote the resulting tableau by $j_c(P)$. Similarly, a *backward slide on P into a cell c* produces a tableau $j^c(P)$ by

    R1 Pick c to be an inner corner of μ.

    R2 **While** c is not an inner corner of λ **do**

        Ra If $c = (i, j)$, then let c' be the cell of $\min\{P_{i+1,j}, P_{i,j+1}\}$.
        Rb Slide $P_{c'}$ into cell c and let $c := c'$. ∎

By way of illustration, let

$$P = \begin{array}{cccc} & & 6 & 8 \\ & 2 & 4 & 5 \quad 9 \\ 1 & 3 & 7 & \end{array}$$

We let a dot indicate the position of the empty cell as we perform a forward slide into $c = (3, 4)$.

$$
\begin{array}{ccc}
& 6 \;\; 8 \\
2 \;\; 4 \;\; 5 \;\; 9 \;, \\
1 \;\; 3 \;\; 7 \;\; \bullet
\end{array}
\quad
\begin{array}{ccc}
& 6 \;\; 8 \\
2 \;\; 4 \;\; 5 \;\; 9 \;, \\
1 \;\; 3 \;\; \bullet \;\; 7
\end{array}
\quad
\begin{array}{ccc}
& 6 \;\; 8 \\
2 \;\; \bullet \;\; 5 \;\; 9 \;, \\
1 \;\; 3 \;\; 4 \;\; 7
\end{array}
\quad
\begin{array}{ccc}
& 6 \;\; 8 \\
\bullet \;\; 2 \;\; 5 \;\; 9 \\
1 \;\; 3 \;\; 4 \;\; 7
\end{array}
$$

So

$$
j_c(P) = \quad
\begin{array}{ccc}
6 \;\; 8 \\
2 \;\; 5 \;\; 9 \\
1 \;\; 3 \;\; 4 \;\; 7
\end{array}
$$

A backward slide into $c = (1, 3)$ looks like the following.

$$
\begin{array}{ccc}
\bullet \;\; 6 \;\; 8 \\
2 \;\; 4 \;\; 5 \;\; 9 \;, \\
1 \;\; 3 \;\; 7
\end{array}
\quad
\begin{array}{ccc}
4 \;\; 6 \;\; 8 \\
2 \;\; \bullet \;\; 5 \;\; 9 \;, \\
1 \;\; 3 \;\; 7
\end{array}
\quad
\begin{array}{ccc}
4 \;\; 6 \;\; 8 \\
2 \;\; 5 \;\; \bullet \;\; 9 \;, \\
1 \;\; 3 \;\; 7
\end{array}
\quad
\begin{array}{ccc}
4 \;\; 6 \;\; 8 \\
2 \;\; 5 \;\; 9 \;\; \bullet \\
1 \;\; 3 \;\; 7
\end{array}
$$

Thus

$$
j^c(P) = \quad
\begin{array}{ccc}
4 \;\; 6 \;\; 8 \\
2 \;\; 5 \;\; 9 \\
1 \;\; 3 \;\; 7
\end{array}
$$

Note that a slide is an invertible operation. Specifically, if c is a cell for a forward slide on P and the cell vacated by the slide is d, then a backward slide into d restores P. In symbols,

$$
j^d j_c(P) = P \tag{3.16}
$$

Similarly

$$
j_c j^d(P) = P \tag{3.17}
$$

if the roles of d and c are reversed.

Of course, we may want to make many slides in succession.

Definition 3.9.3 A sequence of cells $(c_1, c_2, \ldots, c_l)$ is a *slide sequence* for a tableau P if we can legally form $P = P_0, P_1, \ldots, P_l$, where P_i is obtained from P_{i-1} by performing a slide into cell c_i. Partial tableaux P and Q are *equivalent*, written $P \cong Q$, if Q can be obtained from P by some sequence of slides. ∎

This equivalence relation is the same as Knuth equivalence, as the next series of results show.

Proposition 3.9.4 ([Scü 76]) *If P, Q are standard skew tableaux, then*

$$
P \cong Q \Longrightarrow P \stackrel{K}{\cong} Q
$$

Proof. By induction, it suffices to prove the theorem when P and Q differ by a single slide. In fact, if we call the operation in steps Fb or Rb of the slide definition a *move*, then we need to demonstrate the result only when P and Q differ by a move. (The row word of a tableau with a hole in it can still be defined by merely ignoring the hole.)

The conclusion is trivial if the move is horizontal because then $\pi_P = \pi_Q$. If the move is vertical, then we can clearly restrict to the case where P and Q have only two rows. So suppose that x is the element being moved and that

$$
P = \begin{array}{|c|c|c|c|}
\hline
 & R_l & x & R_r \\
\hline
S_l & & \bullet & S_r \\
\hline
\end{array}
$$

$$
Q = \begin{array}{|c|c|c|c|}
\hline
 & R_l & \bullet & R_r \\
\hline
S_l & & x & S_r \\
\hline
\end{array}
$$

where R_l and S_l (respectively, R_r and S_r) are the left (respectively, right) portions of the two rows.

Now induct on the number of elements in P (or Q). If both tableaux consist only of x, then we are done.

Now suppose $|R_r| > |S_r|$. Let y be the rightmost element of R_r and let P', Q' be P, Q, respectively, with y removed. By our assumption P' and Q' are still skew tableau, so applying induction

$$
\pi_P = \pi_{P'} y \overset{K}{\cong} \pi_{Q'} y = \pi_Q
$$

The case $|S_l| > |R_l|$ is handled similarly.

Thus we are reduced to considering $|R_r| = |S_r|$ and $|R_l| = |S_l|$. Say

$$
R_l = x_1 \ldots x_j, \qquad R_r = y_1 \ldots y_k
$$
$$
S_l = z_1 \ldots z_j, \qquad S_r = w_1 \ldots w_k
$$

By induction, we may assume that one of j or k is positive. We'll handle the situation where $j > 0$, leaving the other case to the reader. The following simple lemma will prove convenient.

Lemma 3.9.5 *Suppose* $a_1 < a_2 < \cdots < a_n$.

1. *If* $x < a_1$, *then* $a_1 \ldots a_n x \overset{K}{\cong} a_1 x a_2 \ldots a_n$.

2. *If* $x > a_n$, *then* $x a_1 \ldots a_n \overset{K}{\cong} a_1 \ldots a_{n-1} x a_n$. ∎

Since the rows and columns of P increase, we have $x_1 < z_i$ and $x_1 < w_i$ for all i as well as $x_1 < x$. Thus

$$
\begin{aligned}
\pi_P &= z_1 \ldots z_j w_1 \ldots w_k x_1 \ldots x_j x y_1 \ldots y_k \\
&\overset{K}{\cong} z_1 x_1 z_2 \ldots z_j w_1 \ldots w_k x_2 \ldots x_j x y_1 \ldots y_k \quad \text{(Lemma 3.9.5, part 1)} \\
&\overset{K}{\cong} z_1 x_1 z_2 \ldots z_j x w_1 \ldots w_k x_2 \ldots x_j y_1 \ldots y_k \quad \text{(induction)} \\
&\overset{K}{\cong} z_1 \ldots z_j x w_1 \ldots w_k x_1 \ldots x_j y_1 \ldots y_k \quad \text{(Lemma 3.9.5, part 1)} \\
&= \pi_Q \qquad\qquad\qquad\qquad\qquad\qquad\qquad\qquad\qquad ∎
\end{aligned}
$$

Schützenberger's teasing game can be described in the following manner.

Definition 3.9.6 Given a partial skew tableau P, we play *jeu de taquin* by choosing an arbitrary slide sequence that brings P to normal shape and then applying the slides. The resulting tableau is denoted $j(P)$. ∎

It is not obvious at first blush that $j(P)$ is well defined—i.e., independent of the slide sequence. However, it turns out that we will always get the Robinson-Schensted P-tableau for the row word of P.

Theorem 3.9.7 ([Scü 76]) *Suppose P is a partial skew tableau that is brought to a normal tableaux P' by slides. Then P' is unique—in fact, P' is the insertion tableau for π_P.*

Proof. By the previous proposition, $\pi_P \stackrel{K}{\cong} \pi_{P'}$. Thus by Knuth's theorem on P-equivalence (Theorem 3.6.3), π_P and $\pi_{P'}$ have the same insertion tableau. Finally, Lemma 3.6.5 tells us that the insertion tableau for $\pi_{P'}$ is just P' itself. ∎

We end this section by showing that equivalence and Knuth equivalence are indeed equivalent.

Theorem 3.9.8 ([Scü 76]) *Let P and Q be partial skew tableaux. Then*

$$P \cong Q \iff P \stackrel{K}{\cong} Q$$

Proof. The only-if direction is Proposition 3.9.4. For the other implication, note that since $P \stackrel{K}{\cong} Q$, their row words must have the same P-tableau (Theorem 3.6.3 again). So by the previous theorem, $j(P) = j(Q) = P'$, say. Thus we can take P into Q by performing the slide sequence taking P to P' and then the inverse of the sequence taking Q to P'. Hence $P \cong Q$. ∎

3.10 Dual Equivalence

In the last section we gave a characterization of Knuth equivalence in terms of slides (Theorem 3.9.8). It would be nice to have a corresponding characterization of dual Knuth equivalence, and this was done by Haiman [Hai pr]. Haiman's machinery will also be useful when we prove the Littlewood-Richardson rule in Section 4.9. Before stating Haiman's key definition, we prove a useful result about normal tableaux.

Proposition 3.10.1 *Let P and Q be standard with the same normal shape $\lambda \vdash n$. Then $P \stackrel{K^*}{\cong} Q$.*

Proof. Induct on n, the proposition being trivial for $n \leq 2$. When $n \geq 3$, let c and d be the inner corner cells containing n in P and Q, respectively. There are two cases, depending on the relative positions of c and d.

If $c = d$, then let P' (respectively, Q') be P (respectively, Q) with the n erased. Now P' and Q' have the same shape, so by induction $\pi_{P'} \overset{K^*}{\cong} \pi_{Q'}$. But then we can apply the same sequence of dual Knuth relations to π_P and π_Q, the presence of n being immaterial. Thus $P \overset{K^*}{\cong} Q$ in this case.

If $c \neq d$, then it suffices to show the existence of two dual Knuth equivalent tableaux P' and Q' with n in cells c and d, respectively. (Because then, by what we have shown in the first case, it follows that

$$\pi_P \overset{K^*}{\cong} \pi_{P'} \overset{K^*}{\cong} \pi_{Q'} \overset{K^*}{\cong} \pi_Q$$

and we are done.) Let e be a lowest rightmost cell among all the cells on the boundary of λ between c and d. (The *boundary of* λ is the set of all cells at the end of a row or column of λ.) Schematically, we might have the situation

Now let

$$P'_c = n, \qquad P'_e = n - 2, \quad P'_d = n - 1$$
$$Q'_c = n - 1, \quad Q'_e = n - 2, \quad Q'_d = n$$

and place the numbers $1, 2, \ldots, n - 3$ anywhere as long as they are in the same cells in both P' and Q'. By construction, $\pi_P \overset{K^*}{\cong} \pi_Q$. ∎

The definition of dual equivalence is as follows.

Definition 3.10.2 Partial skew tableaux P and Q are *dual equivalent*, written $P \overset{*}{\cong} Q$, if whenever we apply the same slide sequence to both P and Q, we get resultant tableaux of the same shape. ∎

Note that the empty slide sequence can be applied to any tableau, so $P \overset{*}{\cong} Q$ implies that $\operatorname{sh} P = \operatorname{sh} Q$. The next result is proved directly from our definitions.

Lemma 3.10.3 Let $P \overset{*}{\cong} Q$. If applying the same sequence of slides to both tableaux yield P' and Q', then $P' \overset{*}{\cong} Q'$. ∎

One tableau may be used to determine a sequence of slides for another as follows. Let P and Q have shapes μ/ν and λ/μ, respectively. Then the cells of Q, taken in the order determined by Q's entries, are a sequence of forward slides on P. Let $j_Q(P)$ denote the result of applying this slide sequence to P. Also let $V = v_Q(P)$ stand for the tableau formed by the sequence of cells vacated during the construction of $j_Q(P)$—i.e.,

$$V_{i,j} = k \qquad \text{if } (i,j) \text{ was vacated when filling the cell of } k \in Q \qquad (3.18)$$

Displaying the elements of P as boldface and those of Q in normal type, we can compute an example.

$$
\begin{array}{l}
\mathbf{1}\ \mathbf{2}\ \mathbf{3} \\
\mathbf{4}\ \mathbf{1}\ \mathbf{3} \\
\mathbf{2}
\end{array}
\ ,\quad
\begin{array}{l}
\ \ \mathbf{2}\ \mathbf{3} \\
\mathbf{1}\ \mathbf{4}\ \mathbf{3} \\
\ \ \mathbf{2}
\end{array}
\ ,\quad
\begin{array}{l}
\ \ \mathbf{2}\ \mathbf{3} \\
\ \ \mathbf{4}\ \mathbf{3} \\
\mathbf{1}
\end{array}
\ ,\quad
\begin{array}{l}
\ \ \ \ \mathbf{3} \\
\mathbf{2}\ \mathbf{4} \\
\mathbf{1}
\end{array}
\quad = j_Q(\mathbf{P})
$$

$$
\emptyset,\qquad \mathbf{1},\qquad
\begin{array}{l}
\mathbf{1} \\
\mathbf{2}
\end{array}
\ ,\qquad
\begin{array}{l}
\mathbf{1}\ \mathbf{3} \\
\mathbf{2}
\end{array}
\quad = v_Q(\mathbf{P})
$$

With P and Q as before, the entries of P taken in reverse order define a sequence of backward slides on Q. Performing this sequence gives a tableau denoted $j^P(Q)$. The vacating tableau $V = v^P(Q)$ is still defined by equation (3.18), with P in place of Q. The reader can check that using our example tableaux we obtain $j^P(Q) = v_Q(P)$ and $v^P(Q) = j_Q(P)$. This always happens, although it is not obvious. The interested reader should consult [Hai pr].

From these definitions it is clear that we can generalize equations (3.16) and (3.17). Letting $V = v^P(Q)$ and $W = v_Q(P)$, we have

$$
j^W j_Q(P) = P \text{ and } j_V j^P(Q) = Q \tag{3.19}
$$

To show that dual equivalence and dual Knuth equivalence are really the same, we have to concentrate on small tableaux first.

Definition 3.10.4 A partial tableau P is *miniature* if P has exactly three elements. ∎

The miniature tableaux are used to model the dual Knuth relations of the first and second kinds.

Proposition 3.10.5 *Let P and Q be distinct miniature tableaux of the same shape λ/μ and content. Then*

$$
P \stackrel{K^*}{\cong} Q \iff P \stackrel{*}{\cong} Q
$$

Proof. Without loss of generality, let P and Q be standard.

$\implies$ By induction on the number of slides, it suffices to show the following. Let c be a cell for a slide on P, Q and let P', Q' be the resultant tableaux. Then we must have

$$
\operatorname{sh} P' = \operatorname{sh} Q' \text{ and } P' \stackrel{K^*}{\cong} Q' \tag{3.20}
$$

This is a tedious case-by-case verification. First, we must write down all the skew shapes with 3 cells (up to those translations that do not affect slides so the number of diagrams will be finite). Then we must find the possible tableau pairs for each shape (there will be at most two pairs corresponding to $\stackrel{1^*}{\cong}$ and $\stackrel{2^*}{\cong}$). Finally, all possible slides must be tried on each pair. We leave the details to the reader. However, we will do one of the cases as an illustration.

Suppose that $\lambda = (2,1)$ and $\mu = \emptyset$. Then the only pair of tableaux of this shape is

$$P = \begin{array}{cc} 1 & 2 \\ 3 & \end{array} \quad \text{and} \quad Q = \begin{array}{cc} 1 & 3 \\ 2 & \end{array}$$

or vice versa, and $P \stackrel{1^*}{\cong} Q$. The results of the three possible slides on P, Q are given in the following table, from which it is easy to verify that (3.20) holds.

c :	$(1,3)$	$(2,2)$	$(3,1)$
P' :	$\begin{array}{cc} 1 & 2 \\ 3 & \end{array}$	$\begin{array}{cc} & 2 \\ 1 & 3 \end{array}$	$\begin{array}{c} 2 \\ 1 \\ 3 \end{array}$
Q' :	$\begin{array}{cc} 1 & 3 \\ 2 & \end{array}$	$\begin{array}{cc} & 1 \\ 2 & 3 \end{array}$	$\begin{array}{c} 3 \\ 1 \\ 2 \end{array}$
$\stackrel{i^*}{\cong}$:	$\stackrel{1^*}{\cong}$	$\stackrel{2^*}{\cong}$	$\stackrel{1^*}{\cong}$

$\Longleftarrow$ Let N be a normal standard tableau of shape μ. So $P' = j^N(P)$ and $Q' = j^N(Q)$ are normal miniature tableaux. Now $P \stackrel{*}{\cong} Q$ implies that $\operatorname{sh} P' = \operatorname{sh} Q'$. This hypothesis also guarantees that $v^N(P) = v^N(Q) = V$ say. Applying equation (3.19),

$$j_V(P') = P \neq Q = j_V(Q')$$

which gives $P' \neq Q'$. Thus P' and Q' are distinct miniature tableaux of the same normal shape. The only possibility is, then,

$$\{P', Q'\} = \left\{ \begin{array}{cc} 1 & 2 \\ 3 & \end{array}, \begin{array}{cc} 1 & 3 \\ 2 & \end{array} \right\}$$

Since $P' \stackrel{K^*}{\cong} Q'$, we have, by what was proved in the forward direction,

$$P = j_V(P') \stackrel{K^*}{\cong} j_V(Q') = Q \quad \blacksquare$$

To make it more convenient to talk about miniature subtableaux of a larger tableau, we make the following definition.

Definition 3.10.6 Let P and Q be standard skew tableaux with

$$\operatorname{sh} P = \mu/\nu \vdash m \quad \text{and} \quad \operatorname{sh} Q = \lambda/\mu \vdash n$$

Then $P \cup Q$ denotes the tableau of shape $\lambda/\nu \vdash m + n$ such that

$$(P \cup Q)_c = \left\{ \begin{array}{ll} P_c & \text{if } c \in \mu/\nu \\ Q_c + m & \text{if } c \in \lambda/\mu \end{array} \right. \blacksquare$$

Using the P and Q on page 125, we have

$$P \cup Q = \begin{array}{ccc} 1 & 2 & 3 \\ 4 & 5 & 7 \\ 6 & & \end{array}$$

We need one more lemma before the main theorem of this section.

Lemma 3.10.7 ([Hai pr]) *Let V, W, P and Q be standard skew tableaux with*

$$\operatorname{sh} V = \mu/\nu, \qquad \operatorname{sh} P = \operatorname{sh} Q = \lambda/\mu, \qquad \operatorname{sh} W = \kappa/\lambda$$

Then

$$P \overset{*}{\cong} Q \Longrightarrow V \cup P \cup W \overset{*}{\cong} V \cup Q \cup W$$

Proof. Consider what happens when performing a single backward slide on $V \cup P \cup W$, say into cell c. Because of the relative order of the elements in the V, P, and W portions of the tableau, the slide can be broken up into three parts. First of all, the slide travels through V, creating a new tableaux $V' = j_c(V)$ and vacating some inner corner d of μ. Then P becomes $P' = j_d(P)$, vacating cell e, and finally W is transformed into $W' = j_e(W)$. Thus $j_c(V \cup P \cup W) = V' \cup P' \cup W'$.

Now perform the same slide on $V \cup Q \cup W$. Tableau V is replaced by $j_c(V) = V'$, vacating d. If $Q' = j_d(Q)$, then, since $P \overset{*}{\cong} Q$, we have $\operatorname{sh} P' = \operatorname{sh} Q'$. So e is vacated as before, and W becomes W'. Thus $j_c(V \cup Q \cup W) = V' \cup Q' \cup W'$ with $P' \overset{*}{\cong} Q'$ by Lemma 3.10.3.

Now the preceding also holds, *mutatis mutandis*, to forward slides. Hence applying the same slide to both $V \cup P \cup W$ and $V \cup P \cup W$ yields tableaux of the same shape that still satisfy the hypotheses of the lemma. By induction, we are done. ∎

We can now show that Proposition 3.10.5 actually holds for all pairs of tableaux.

Theorem 3.10.8 ([Hai pr]) *Let P and Q be standard tableaux of the same shape λ/μ. Then*

$$P \overset{K^*}{\cong} Q \iff P \overset{*}{\cong} Q$$

Proof. $\Longrightarrow$ We need to consider only the case where P and Q differ by a single dual Knuth relation, say the first (the second is similar). Now Q is obtained from P by switching $k+1$ and $k+2$ for some k. So

$$P = V \cup P' \cup W \text{ and } Q = V \cup Q' \cup W$$

where P' and Q' are the miniature subtableaux of P and Q, respectively, that contain k, $k+1$, and $k+2$. By hypothesis, $P' \overset{1^*}{\cong} Q'$, which implies $P' \overset{*}{\cong} Q'$ by Proposition 3.10.5. But then the lemma just proved applies to show that $P \overset{*}{\cong} Q$.

$\Longleftarrow$ Let tableau N be of normal shape μ. Let

$$P' = j^N(P) \text{ and } Q' = j^N(Q)$$

Since $P \stackrel{*}{\cong} Q$, we have $P' \stackrel{*}{\cong} Q'$ (Lemma 3.10.3) and $v^N(P) = v^N(Q) = V$ for some tableau V. Thus, in particular, $\operatorname{sh} P' = \operatorname{sh} Q'$, so that P' and Q' are dual Knuth equivalent by Proposition 3.10.1. Now, by definition, we have a sequence of dual Knuth relations

$$P' = P_1 \stackrel{i'^*}{\cong} P_2 \stackrel{j'^*}{\cong} \cdots \stackrel{l'^*}{\cong} P_k = Q'$$

where $i', j', \ldots, l' \in \{1, 2\}$. Hence the proof of the forward direction of Proposition 3.10.5 and equation (3.19) show that

$$P = j_V(P') \stackrel{i^*}{\cong} j_V(P_2) \stackrel{j^*}{\cong} \cdots \stackrel{l^*}{\cong} j_V(Q') = Q$$

for some $i, j, \ldots, l \in \{1, 2\}$. This finishes the proof of the theorem. $\blacksquare$

3.11   Evacuation

We now return to our project of determining the effect that a reflection or rotation of the permutation matrix for π has on the tableaux $P(\pi)$ and $Q(\pi)$. We have already seen what happens when π is replaced by π^{-1} (Theorem 3.8.6). Also, Theorem 3.4.3 tells us what the P-tableau of π^r looks like. Since these two operations correspond to reflections that generate the dihedral group of the square, we will be done as soon as we determine $Q(\pi^r)$. To do this, another concept called evacuation [Scü 63] is needed.

Definition 3.11.1 Let Q be a partial skew tableau and let m be the minimal element of Q. Then the *delta operator applied to Q* yields a new tableau, ΔQ, defined as follows.

D1 Erase m from its cell, c, in Q.

D2 Perform the slide j^c on the resultant tableau.

If Q is standard with n elements, then the *evacuation tableau for Q* is the vacating tableau $V = \operatorname{ev} Q$ for the sequence

$$Q, \Delta Q, \Delta^2 Q, \ldots, \Delta^n Q$$

That is,

$$V_d = n - i \quad \text{if cell } d \text{ was vacated when passing from } \Delta^i Q \text{ to } \Delta^{i+1} Q \quad \blacksquare$$

Taking

$$Q = \begin{array}{l} 1\ 3\ 4\ 7 \\ 2\ 5 \\ 6 \end{array}$$

we compute ev Q as follows, using dots as placeholders for cells not yet filled.

$$\Delta^i Q : \begin{array}{llllllll} 1\ 3\ 4\ 7, & 2\ 3\ 4\ 7, & 3\ 4\ 7, & 4\ 7, & 5\ 7, & 6\ 7, & 7 \\ 2\ 5 & 5 & 5 & 5 & 6 & & \\ 6 & 6 & 6 & 6 & & & \end{array}$$

$$\text{ev}\, Q : \begin{array}{llllllll} \bullet\ \bullet\ \bullet\ \bullet, & \bullet\ \bullet\ \bullet\ \bullet, & \bullet\ \bullet\ \bullet\ 6, & \bullet\ \bullet\ 5\ 6, & \bullet\ \bullet\ 5\ 6, & \bullet\ \bullet\ 5\ 6, & \bullet\ 2\ 5\ 6 \\ \bullet\ \bullet & \bullet\ 7 & \bullet\ 7 & \bullet\ 7 & \bullet\ 7 & 3\ 7 & 3\ 7 \\ \bullet & \bullet & \bullet & \bullet & 4 & 4 & 4 \end{array}$$

Completing the last slide we obtain

$$\text{ev}\, Q = \begin{array}{llll} 1 & 2 & 5 & 6 \\ 3 & 7 & & \\ 4 & & & \end{array}$$

Note that Q was the Q-tableau of our example permutation $\pi = 4\ 2\ 3\ 6\ 5\ 1\ 7$ in (3.7). The reader who wishes to anticipate Theorem 3.11.4 should now compute $Q(\pi^r)$ and compare the results.

Since the delta operator is defined in terms of slides, it is not surprising that it commutes with jeu de taquin.

Lemma 3.11.2 *Let Q be any skew partial tableau; then*

$$j\Delta(Q) = \Delta j(Q)$$

Proof. Let P be Q with its minimum element m erased from cell c. We write this as $P = Q - \{m\}$. Then

$$j\Delta(Q) = j(P)$$

by the definition of Δ and the unicity of the j operator (Theorem 3.9.7).

We now show that any backward slide on Q can be mimicked by a slide on P. Let $Q' = j^d(Q)$ for some cell d. There are two cases. If d is not vertically or horizontally adjacent to c, then it is legal to form $P' = j^d(P)$. Since m is involved in neither slide, we again have that $P' = Q' - \{m\}$.

If d is adjacent to c, then the first move of $j^d(Q)$ will put m in cell d and then proceed as a slide into c. Thus letting $P' = j^c(P)$ will preserve the relationship between P' and Q' as before.

Continuing in this manner, when we obtain $j(Q)$, we will have a corresponding P'', which is just $j(Q)$ with m erased from the $(1,1)$ cell. Thus

$$\Delta j(Q) = j^{(1,1)}(P'') = j(P) = j\Delta(Q) \quad \blacksquare$$

As a corollary, we obtain our first relationship between the jeu de taquin and the Q-tableau of the Robinson-Schensted algorithm.

Proposition 3.11.3 ([Scü 63]) *Suppose* $\pi = x_1 x_2 \ldots x_n \in \mathcal{S}_n$ *and let*

$$\overline{\pi} = \begin{array}{cccc} 2 & 3 & \cdots & n \\ x_2 & x_3 & \cdots & x_n \end{array}$$

Then

$$Q(\overline{\pi}) = \Delta Q(\pi)$$

Proof. Consider $\sigma = \pi^{-1}$ and $\overline{\sigma} = \overline{\pi}^{-1}$. Note that the lower line of $\overline{\sigma}$ is obtained from the lower line of σ by deleting the minimum element, 1.

By Theorem 3.8.6, it suffices to show that

$$P(\overline{\sigma}) = \Delta P(\sigma)$$

If we view σ and $\overline{\sigma}$ as anti-diagonal strip tableaux, then $\overline{\sigma} \cong \Delta\sigma$. Hence by Theorem 3.9.7 and Lemma 3.11.2,

$$P(\overline{\sigma}) = P(\Delta\sigma) = \Delta P(\sigma) \ \blacksquare$$

Finally, we can complete our characterization of the image of π^r under the Robinson-Schensted map.

Theorem 3.11.4 ([Scü 63]) *If* $\pi \in \mathcal{S}_n$, *then*

$$Q(\pi^r) = \mathrm{ev}\, Q(\pi)^t$$

Proof. Let $\pi, \overline{\pi}$ be as in the previous proposition with

$$\overline{\pi}^r = \begin{array}{cccc} 1 & 2 & \cdots & n-1 \\ x_n & x_{n-1} & \cdots & x_2 \end{array}$$

Induct on n. Now

$$\begin{aligned} Q(\pi^r) - \{n\} &= Q(\overline{\pi}^r) && (\overline{\pi}^r = x_n \ldots x_2) \\ &= \mathrm{ev}\, Q(\overline{\pi})^t && \text{(induction)} \\ &= \mathrm{ev}\, Q(\pi)^t - \{n\} && \text{(Proposition 3.11.3)} \end{aligned}$$

Thus we need to show only that n occupies the same cell in both $Q(\pi^r)$ and $\mathrm{ev}\, Q(\pi)^t$. Let $\mathrm{sh}\, Q(\pi) = \lambda$ and $\mathrm{sh}\, Q(\overline{\pi}) = \overline{\lambda}$. By Theorems 3.3.1 and 3.4.3 we have $\mathrm{sh}\, Q(\pi^r) = \lambda^t$ and $\mathrm{sh}\, Q(\overline{\pi}^r) = \overline{\lambda}^t$. Hence

$$\begin{aligned} \text{cell of } n \text{ in } Q(\pi^r) &= \lambda^t/\overline{\lambda}^t && (\overline{\pi}^r = x_n \ldots x_2) \\ &= (\lambda/\overline{\lambda})^t \\ &= (\text{cell of } n \text{ in } \mathrm{ev}\, Q(\pi))^t && \text{(Proposition 3.11.3)} \ \blacksquare \end{aligned}$$

3.12 Exercises

1. A poset τ is a *rooted tree* if it has a unique minimal element and its Hasse diagram has no cycles. If $v \in \tau$, then define its *hook* to be

$$H_v = \{w \in \tau \mid w \geq v\}$$

with corresponding *hooklength* $h_v = |H_v|$. Show in two ways, inductively and probabilistically, that if τ has n nodes, then the number of natural labelings of τ (see Definition 4.2.5) is

$$f^\tau = \frac{n!}{\prod_{v \in \tau} h_v}$$

2. A *two-person ballot sequence* is a permutation $\pi = x_1 x_2 \ldots x_{2n}$ of n ones and n twos such that, for any prefix $\pi_k = x_1 x_2 \ldots x_k$, the number of ones in π_k is at least as great as the number of twos. The nth *Catalan number*, C_n, is the number of such sequences.

   a. Prove the recurrence

   $$C_{n+1} = C_n C_0 + C_{n-1} C_1 + \cdots + C_0 C_n$$

   for $n \geq 0$.

   b. The Catalan numbers also count the following sets. Show this in two ways: by verifying that the recurrence is satisfied and by giving a bijection with a set of objects already known to be counted by the C_n.

      i. Standard tableaux of shape (n, n).
      ii. Permutations $\pi \in \mathcal{S}_n$ with longest decreasing subsequence of length at most two.
      iii. Sequences of positive integers

      $$1 \leq a_1 \leq a_2 \leq \cdots \leq a_n$$

      such that $a_i \leq i$ for all i.
      iv. Binary trees on n nodes.
      v. Pairings of $2n$ labeled points on a circle with chords that do not cross.
      vi. Triangulations of a convex $(n+2)$-gon by diagonals.
      vii. Lattice paths of the type considered in Section 4.5 from $(0,0)$ to (n, n) that stay weakly below the diagonal $y = x$.
      viii. Non-crossing partitions of $\{1, 2, \ldots, n\}$—i.e., ways of writing

      $$\{1, 2, \ldots, n\} = B_1 \uplus B_2 \uplus \cdots \uplus B_k$$

      such that $a, c \in B_i$ and $b, d \in B_j$ with $a < b < c < d$ implies $i = j$.

c. Prove, by using results of this chapter, that

$$C_n = \frac{1}{n+1}\binom{2n}{n}$$

3. If $\lambda = (\lambda_1, \lambda_2, \ldots, \lambda_l) \vdash n$, then derive the following formulae

$$
\begin{aligned}
f^\lambda &= n! \frac{\prod_{i<j}(\lambda_i - \lambda_j - i + j)}{\prod_i (\lambda_i - i + l)!} \\
&= n! \frac{\prod_{i<j}(h_{i,1} - h_{j,1})}{\prod_i h_{i,1}!}
\end{aligned}
$$

4. A partition $\lambda = (\lambda_1, \lambda_2, \ldots, \lambda_l) \vdash n$ is *strict* if $\lambda_1 > \lambda_2 > \cdots > \lambda_l$. Given any strict partition, the associated *shifted shape* λ^* indents row i of the normal shape so that it starts on the diagonal square (i, i). For example, the shifted shape of $\lambda = (4, 3, 1)$ is

$$
\lambda^* = \quad
\begin{matrix}
\bullet & \bullet & \bullet & \bullet \\
 & \bullet & \bullet & \bullet \\
 & & \bullet &
\end{matrix}
$$

The *shifted hook* of $(i, j) \in \lambda^*$ is

$$H_{i,j}^* = \{(i, j') \mid j' \geq j\} \cup \{(i', j) \mid i' \geq i\} \cup \{(j+1, j') \mid j' \geq j+1\}$$

with *shifted hooklength* $h_{i,j}^* = |H_{i,j}^*|$. For λ^* as before, the shifted hooklengths are

$$
\begin{matrix}
7 & 5 & 4 & 2 \\
  & 4 & 3 & 1 \\
  &   & 1 &
\end{matrix}
$$

a. Show that the $h_{i,j}^*$ can be obtained as ordinary hooklengths in a left-justified diagram obtained by pasting together λ^* with its transpose.

b. Defining standard shifted tableaux in the obvious way, show that the number of such arrays is given by

$$
\begin{aligned}
f^{\lambda^*} &= n! \frac{\prod_{i<j}(\lambda_i - \lambda_j)}{\prod_i \lambda_i! \prod_{i<j}(\lambda_i + \lambda_j)} \\
&= \frac{n!}{\prod_{i<j}(\lambda_i + \lambda_j)} \det \frac{1}{(\lambda_i - l + j)!} \\
&= \frac{n!}{\prod_{(i,j)\in\lambda^*} h_{i,j}^*}
\end{aligned}
$$

5. Prove Lemma 3.4.1.

6. Let P be a partial tableau with $x, y \notin P$. Suppose the paths of insertion for $r_x(P) = P'$ and $r_y(P') = P''$ are

$$(1, j_1), \ldots, (r, j_r) \quad \text{and} \quad (1, j'_1), \ldots, (r', j'_{r'})$$

respectively. Show the following.

   a. If $x < y$, then $j_i < j'_i$ for all $i \leq r'$ and $r' \leq r$; i.e., the path for r_y lies strictly to the right of the path for r_x.

   b. If $x > y$, then $j_i \geq j'_i$ for all $i \leq r$ and $r' > r$; i.e., the path for r_y lies weakly to the left of the path for r_x.

7.   a. If the Robinson-Schensted algorithm is only used to find the length of the longest increasing subsequence of $\pi \in S_n$, find its computational complexity in the worst case.

   b. Use the algorithm to find all increasing subsequences of π of maximum length.

   c. Use Viennot's construction to find all increasing subsequences of π of maximum length.

8.   a. Use the results of this chapter to prove the Erdös-Szekeres theorem: given any $\pi \in S_{nm+1}$, then π contains either an increasing subsequence of length $n + 1$ or a decreasing subsequence of length $m + 1$.

   b. This theorem can be made into a game as follows. Player A picks $x_1 \in S = \{1, 2, \ldots nm + 1\}$. Then player B picks $x_2 \in S$ with $x_2 \neq x_1$. The players continue to alternate picking distinct elements of S until the sequence $x_1 x_2 \ldots$ contains an increasing subsequence of length $n + 1$ or a decreasing subsequence of length $m + 1$. In the achievement (respectively, avoidance) version of the game, the last player to move wins (respectively, loses). With m arbitrary, find winning strategies for $n \leq 2$ in the achievement game and for $n \leq 1$ in the avoidance game. It is an open problem to find a strategy in general.

9. Prove Lemma 3.6.5 and describe the Q-tableau of π_P.

10. Let $\pi = x_1 \ldots x_n$ be a partial permutation. The *Greene invariant of* π is the sequence of increasing subsequence lengths

$$i(\pi) = (i_1(\pi), i_2(\pi), \ldots, i_n(\pi))$$

If $\pi \in S_n$, then let
$$\pi^{(j)} = x_1 \ldots x_j$$

and

$$\pi_{(j)} = \text{the subsequence of } \pi \text{ containing all elements } \leq j$$

a. Show that $P(\pi) = P(\sigma)$ if and only if $i(\pi_{(j)}) = i(\sigma_{(j)})$ for all j by a direct argument.

b. Show that $Q(\pi) = Q(\sigma)$ if and only if $i(\pi^{(j)}) = i(\sigma^{(j)})$ for all j by a direct argument.

c. Show that (a) and (b) are equivalent statements.

11. Recall that an involution is a map π from a set to itself such that π^2 is the identity. Prove the following facts about involutions.

   a. A permutation π is an involution if and only if $P(\pi) = Q(\pi)$. Thus there is a bijection between involutions in $\mathcal{S}_n$ and standard tableaux with n elements.

   b. The number of fixed-points in an involution π is the number of columns of odd length in $P(\pi)$.

   c. [Frobenius-Schur] The number of involutions in $\mathcal{S}_n$ is given by $\sum_{\lambda \vdash n} f^\lambda$.

   d. We have
   $$1 \cdot 3 \cdot 5 \cdots (2n-1) = \sum_{\substack{\lambda \vdash 2n \\ \lambda' \text{even}}} f^\lambda$$
   where λ' even means that the conjugate of λ has only even parts.

12. Let $\pi = x_1 x_2 \ldots x_n$ be any sequence of positive integers, possibly with repetitions. Define the *P-tableau of* π, $P(\pi)$, to be

$$r_{x_n} r_{x_{n-1}} \cdots r_{x_1}(\emptyset)$$

where the row insertion operator is defined as usual.

   a. Show that $P(\pi)$ is a semistandard tableau.

   b. Find and prove the semistandard analogs of each of the following results in the text. Some of the definitions may have to be modified. In each case you should try to obtain the new result as a corollary of the old one rather than rewriting the old proof to account for repetitions.

      i. Theorem 3.6.3
      ii. Theorem 3.7.3
      iii. Theorem 3.9.7

13. Let P be a standard shifted tableau. Let P^2 denote the left-justified semistandard tableau gotten by pasting together P and P^t, as was done for shapes in Exercise 4.

Also define the insertion operation $i_x(P)$ by row inserting x into P until an element comes to rest at the end of a row (and insertion terminates) or a diagonal element in cell (d, d) is displaced. In the latter case $P_{d,d}$ bumps into *column* $d+1$ and bumping continues by columns until some element comes to rest at the end of a column.

a. Show that $(i_x P)^2 = c_x r_x(P^2)$, where the column insertion operator is modified so that x displaces the element of the first column greater than *or equal to* x, and so on.

b. Find and prove the shifted analogs of each of the following results in the text. Some of the definitions may have to be modified. In each case you should try to obtain the new result as a corollary of the old one rather than rewriting the old proof to account for the shifted tableaux.

   i. Theorem 3.6.3

   ii. Theorem 3.7.3

   iii. Theorem 3.9.7

14. Reprove Proposition 3.10.1 using the Robinson-Schensted algorithm.

15. Consider the dihedral group D_4. Then $g \in D_4$ acts on $\pi \in S_n$ by applying the symmetry of the square corresponding to g to the permutation matrix corresponding to π. For every $g \in D_4$, describe $P(g\pi)$ and $Q(g\pi)$ in terms of $P(\pi)$ and $Q(\pi)$. Be sure your description is as simple as possible.

16. Let A be a partially ordered set. A *lower order ideal of P* is $L \subseteq P$ such that $x \in L$ implies that $y \in L$ for all $y \leq x$. *Upper order ideals* U are defined by the reverse inequality. If $P = L \uplus U$, where L and U are lower and upper order ideals, respectively, then define analogs of the jeu de taquin and vacating tableaux on natural labelings of P (see Definition 4.2.5) such that

$$j^L(U) = v_U(L) \text{ and } v^L(U) = j_U(L)$$

Hint: Show that

$$j^L(U) \uplus v^L(U) = j_U(L) \uplus v_U(L)$$

by expressing each slide as a composition of operators that switch elements i and $i+1$ in a labeling as long as it remains natural.

17. Suppose $\pi = x_1 x_2 \ldots x_n$ is a permutation such that $P = P(\pi)$ has rectangular shape. Let the *complement of π* be

$$\pi^c = y_1 y_2 \ldots y_n$$

where $y_i = n + 1 - x_i$ for all i. Also define the *complement of a rectangular standard tableau P* with n entries to be the array obtained by replacing $P_{i,j}$ with $n + 1 - P_{i,j}$ for all (i, j) and then rotating the result 180°. Show that

$$P(\pi^c) = (P^c)^t$$

18. The *reverse delta operator*, Δ', is the same as the delta operator of Definition 3.11.1, with m being replaced by the maximal element of Q and j^c replaced by j_c. The *reverse evacuation tableau*, $\mathrm{ev}'\, Q$, is the vacating tableau for the sequence

$$Q, \Delta'Q, \Delta'^2 Q, \ldots, \Delta'^n Q$$

Prove the following.

a. Evacuation and reverse evacuation are involutions—i.e.,

$$\mathrm{ev}\,\mathrm{ev}\, Q = \mathrm{ev}'\,\mathrm{ev}'\, Q = Q$$

b. If Q has rectangular shape, then

$$\mathrm{ev}'\,\mathrm{ev}\, Q = Q$$

Chapter 4

Symmetric Functions

We have seen how some results about representations of $\mathcal{S}_n$ can be proved either by using general facts from representation theory or combinatorially. There is a third approach using symmetric functions, which is our focus in this chapter.

After giving some general background on formal power series, we derive the hook generating function for semistandard tableaux (generalizing the hook formula, Theorem 3.1.2). The method of proof is a beautiful algorithmic bijection due to Hillman and Grassl [H-G 76].

Next, the symmetric functions themselves are introduced along with the all-important Schur functions, s_λ. The Jacobi-Trudi determinants give alternate expressions for s_λ analogous to the determinantal form for f^λ. The lattice-path techniques of Gessel-Viennot [Ges um, G-V 85, G-V ip] provide a combinatorial proof. Other definitions of the Schur function as a quotient of alternates or as the cycle indicator for the irreducible characters of $\mathcal{S}_n$ are presented. The latter brings in the characteristic map, which is an isomorphism between the algebra of symmetric functions and the algebra of class functions on the symmetric group (which has the irreducible characters as a basis). Knuth's generalization of the Robinson-Schensted map [Knu 70] completes our survey of generating function analogs for results from Chapter 3.

We end by coming full circle with two applications of symmetric functions to representation theory. The first is the Littlewood-Richardson rule [L-R 34], which decomposes a tensor product into irreducibles by looking at the corresponding product of Schur functions. We present a proof based on the jeu de taquin and dual equivalence. The second is a theorem of Murnaghan-Nakayama [Nak 40, Mur 37] which gives an algorithm for computing the irreducible characters. Its proof involves much of the machinery that has been introduced previously.

Those who would like a more extensive and algebraic treatment of symmetric functions should consult Macdonald's book [Mac 79]

4.1   Introduction to Generating Functions

We start with the most basic definition.

Definition 4.1.1 Given a sequence $(a_n)_{n \geq 0} = a_0, a_1, a_2, \ldots$ of complex numbers, the corresponding *generating function* is the power series

$$f(x) = \sum_{n \geq 0} a_n x^n$$

If the a_n enumerate some set of combinatorial objects, then we say that $f(x)$ is the generating function for those objects. We also write

$$[x^n] f(x) = \text{ the coefficient of } x^n \text{ in } f(x) = a_n \blacksquare$$

For example, if

$$a_n = \text{ the number of } n\text{-element subsets of } \{1, 2, 3\}$$

then

$$f(x) = \sum_{n \geq 0} \binom{3}{n} x^n = (1 + x)^3$$

is the generating function for subsets of $\{1, 2, 3\}$. Equivalently,

$$[x^n](1 + x)^3 = \binom{3}{n}$$

It may seem surprising, but to obtain information about a sequence it is often easier to manipulate its generating function. In particular, as we will see shortly, sometimes there is no known simple expression for a_n and yet $f(x)$ is easy to compute. Extensive discussions of generating function techniques can be found in the texts of Goulden and Jackson [G-J 83], Stanley [Stn 86], and Wilf [Wil 90].

Note that all our power series are members of the *formal power series ring*,

$$\mathbf{C}[[x]] = \{\sum_{n \geq 0} a_n x^n \mid a_n \in \mathbf{C} \text{ for all } n\}$$

$\mathbf{C}[[x]]$ is a ring with the usual operations of addition and multiplication. The adjective *formal* refers to the fact that convergence questions are immaterial, since we will never substitute a value for x. The variable and its powers are merely being used to keep track of the coefficients.

We need techniques for deriving generating functions. The basic counting rules for sets state that the English words *or* and *and* are the equivalent of the mathematical operations $+$ and $\times$. Formally, we have the following.

Proposition 4.1.2 *Let S and T be finite sets.*

1. *If $S \cap T = \emptyset$, then*

$$|S \uplus T| = |S| + |T|$$

2. *If S and T are arbitrary, then*

$$|S \times T| = |S| \cdot |T| \quad \blacksquare$$

This proposition has an analog for generating functions (see Proposition 4.1.6). First, however, let us see how these rules can be used informally to compute a few examples.

The basic method for finding the generating function for a given sequence $(a_n)_{n \geq 0}$ is as follows:

1. Find a set S with a parameter such that the number of elements of S whose parameter equals n is a_n.

2. Express the elements of S in terms of *or, and,* and the parameter.

3. Translate this expression into a generating function using $+$, $\times$, and x^n.

In our previous example, $a_n = \binom{3}{n}$, so we can take

$$S = \text{all subsets } T \text{ of } \{1, 2, 3\}$$

The parameter that will produce the sequence is

$$n = n(T) = \text{the number of elements in } T$$

Now we can express any such subset as

$$T = (1 \notin T \text{ or } 1 \in T) \quad \text{and} \quad (2 \notin T \text{ or } 2 \in T) \quad \text{and} \quad (3 \notin T \text{ or } 3 \in T)$$

Finally, translate this expression into a generating function. Remember that the n in x^n is the number of elements in T, so the statements $i \notin T$ and $i \in T$ become x^0 and x^1, respectively. So

$$f(x) = (x^0 + x^1) \cdot (x^0 + x^1) \cdot (x^0 + x^1) = (1 + x)^3$$

as before.

For a more substantial example of the method, let's find the generating function

$$\sum_{n \geq 0} p(n) x^n$$

where $p(n)$ is the number of partitions of n. Here, S is all partitions $\lambda = (1^{m_1}, 2^{m_2}, \ldots)$ and $n = |\lambda|$ is the sum of the parts. We have

$$
\begin{aligned}
\lambda \; = \; & (1^0 \in \lambda \text{ or } 1^1 \in \lambda \text{ or } 1^2 \in \lambda \text{ or } \cdots) \\
& \text{and} \quad (2^0 \in \lambda \text{ or } 2^1 \in \lambda \text{ or } 2^2 \in \lambda \text{ or } \cdots) \\
& \text{and} \quad (3^0 \in \lambda \text{ or } 3^1 \in \lambda \text{ or } 3^2 \in \lambda \text{ or } \cdots) \quad \text{and} \cdots
\end{aligned}
$$

which translates as

$$f(x) = (x^0 + x^1 + x^{1+1} + \cdots)(x^0 + x^2 + x^{2+2} + \cdots)(x^0 + x^3 + x^{3+3} + \cdots) \cdots \quad (4.1)$$

Thus we have proved a famous theorem of Euler [Eul 48].

Theorem 4.1.3 *The generating function for partitions is*

$$\sum_{n \geq 0} p(n)x^n = \frac{1}{1-x} \frac{1}{1-x^2} \frac{1}{1-x^3} \cdots \quad \blacksquare \qquad (4.2)$$

Several remarks about this result are in order. Despite the simplicity of this generating function, there is no known closed-form formula for $p(n)$ itself. (However, there is an expression for $p(n)$ as a sum due to Hardy, Ramanujan and Rademacher, see Theorem 5.1 on page 69 of Andrews [And 76].) This illustrates the power of our approach.

Also, the reader should be suspicious of infinite products such as (4.2). Are they really well-defined elements of $\mathbf{C}[[x]]$? To see what can go wrong, try to find the coefficient of x in $\prod_{i=1}^{\infty}(1 + x)$. To deal with this problem, we need some definitions. Consider $f(x), f_1(x), f_2(x), \ldots \in \mathbf{C}[[x]]$. Then we write $\prod_{i \geq 1} f_i(x) = f(x)$ and say the product *converges* to $f(x)$ if, for every n,

$$[x^n]f(x) = [x^n] \prod_{i=1}^{N} f_i(x)$$

whenever N is sufficiently large. Of course, how large N needs to be depends on n.

A convenient condition for convergence is expressed in terms of the *degree* of $f(x) \in \mathbf{C}[[x]]$, where

$$\deg f(x) = \text{ smallest } n \text{ such that } x^n \text{ has nonzero coefficient in } f(x).$$

For example,

$$\deg(x^2 + x^3 + x^4 + \cdots) = 2$$

The following proposition is not hard to prove and is left to the reader.

Proposition 4.1.4 *If $f_i(x) \in \mathbf{C}[[x]]$ for $i \geq 1$ and $\lim_{i \to \infty} \deg(f_i(x) - 1) = \infty$, then $\prod_{i \geq 1} f_i(x)$ converges.* $\blacksquare$

Note that this shows that the right-hand side of equation (4.2) makes sense, since there $\deg(f_i(x) - 1) = i$.

By carefully examining the derivation of Euler's result, we see that the term $\frac{1}{1-x^i}$ in the product counts the occurrences of i in the partition λ. Thus we can automatically construct other generating functions. For example, if we let

$$p_o(n) = \text{ the number of } \lambda \vdash n \text{ with all parts odd}$$

then

$$\sum_{n \geq 0} p_o(n) x^n = \prod_{i \geq 1} \frac{1}{1 - x^{2i-1}}$$

We can also keep track of the number of parts. To illustrate, consider

$$p_d(n) = \text{ the number of } \lambda \vdash n \text{ with all parts distinct}$$

—i.e., no part of λ appears more than once. Thus the only possibilities for a part i are $i^0 \in \lambda$ or $i^1 \in \lambda$. This amounts to cutting off the generating function in (4.1) after the first two terms, so

$$\sum_{n \geq 0} p_d(n) x^n = \prod_{i \geq 1} (1 + x^i)$$

As a final demonstration of the utility of generating functions, we use them to derive another theorem of Euler [Eul 48] .

Theorem 4.1.5 *For all n, $p_d(n) = p_o(n)$.*

Proof. It suffices to show that $p_d(n)$ and $p_o(n)$ have the same generating function. But

$$\begin{aligned}
\prod_{i \geq 1} (1 + x^i) &= \prod_{i \geq 1} (1 + x^i) \prod_{i \geq 1} \frac{1 - x^i}{1 - x^i} \\
&= \prod_{i \geq 1} \frac{1 - x^{2i}}{1 - x^i} \\
&= \prod_{i \geq 1} \frac{1}{1 - x^{2i-1}} \quad \blacksquare
\end{aligned}$$

It is high time to make more rigorous the steps used to derive generating functions. The crucial definition is as follows. Let S be a set, then a *weighting of S* is a function

$$\text{wt} : S \to \mathbf{C}[[x]]$$

If $s \in S$, then we usually let $\text{wt } s = x^n$ for some n, what we were calling a parameter earlier. The associated *weight generating function* is

$$f_S(x) = \sum_{s \in S} \text{wt } s$$

It is a well-defined element of $\mathbf{C}[[x]]$ as long as the number of $s \in S$ with $\deg \text{wt } s = n$ is finite for every n. To redo our example for the partition function, let S be all partitions, and if $\lambda \in S$, then define

$$\text{wt } \lambda = x^{|\lambda|}$$

This gives the weight generating function

$$
\begin{aligned}
f_S(x) &= \sum_{\lambda \in S} x^{|\lambda|} \\
&= \sum_{n \geq 0} \sum_{\lambda \vdash n} x^n \\
&= \sum_{n \geq 0} p(n) x^n
\end{aligned}
$$

which is exactly what we wish to evaluate.

In order to manipulate weights, we need the corresponding and-or rules.

Proposition 4.1.6 *Let S and T be weighted sets.*

1. *If $S \cap T = \emptyset$, then*

$$
f_{S \uplus T}(x) = f_S(x) + f_T(x)
$$

2. *Let S and T be arbitrary and weight $S \times T$ by $\operatorname{wt}(s,t) = \operatorname{wt} s \operatorname{wt} t$. Then*

$$
f_{S \times T} = f_S(x) \cdot f_T(x)
$$

Proof. 1. If S and T do not intersect, then

$$
\begin{aligned}
f_{S \uplus T}(x) &= \sum_{s \in S \uplus T} \operatorname{wt} s \\
&= \sum_{s \in S} \operatorname{wt} s + \sum_{s \in T} \operatorname{wt} s \\
&= f_S(x) + f_T(x)
\end{aligned}
$$

2. For any two sets S, T, we have

$$
\begin{aligned}
f_{S \times T}(x) &= \sum_{(s,t) \in S \times T} \operatorname{wt}(s,t) \\
&= \sum_{\substack{s \in S \\ t \in T}} \operatorname{wt} s \operatorname{wt} t \\
&= \sum_{s \in S} \operatorname{wt} s \sum_{t \in T} \operatorname{wt} t \\
&= f_S(x) f_T(x) \quad \blacksquare
\end{aligned}
$$

Under suitable convergence conditions, this result can be extended to infinite sums and products. Returning to the partition example:

$$
\begin{aligned}
S &= \{\lambda = (1^{m_1}, 2^{m_2}, \ldots) \mid m_i \geq 0\} \\
&= (1^0, 1^1, 1^2, \ldots) \amalg (2^0, 2^1, 2^2, \ldots) \amalg \cdots \\
&= (\{1^0\} \uplus \{1^1\} \uplus \{1^2\} \uplus \cdots) \amalg (\{2^0\} \uplus \{2^1\} \uplus \{2^2\} \uplus \cdots) \amalg \cdots
\end{aligned}
$$

where $\amalg$ rather than $\times$ is being used, since one is allowed to take only a finite number of components i^{m_i} such that $m_i \neq 0$. So

$$
f_S(x) = (f_{\{1^0\}} + f_{\{1^1\}} + f_{\{1^2\}} + \cdots)(f_{\{2^0\}} + f_{\{2^1\}} + f_{\{2^2\}} + \cdots) \cdots
$$

Since $f_{\{\lambda\}}(x) = \operatorname{wt} \lambda = x^{|\lambda|}$, we recover (4.1) as desired.

4.2   The Hillman-Grassl Algorithm

Just as we were able to enumerate standard λ-tableaux in Chapter 3, we wish to count the semistandard variety of given shape. However, since there are now an infinite number of such arrays, we have to use generating functions. One approach is to sum up the parts, as we do with partitions. This leads to a beautiful hook generating function that was first discovered by Stanley [Stn 71]. The combinatorial proof given next is due to Hillman and Grassl [H-G 76].

Fix a shape λ and let $ss_\lambda(n)$ be the number of semistandard λ-tableaux T such that $\sum_{(i,j)\in\lambda} T_{i,j} = n$. Since all entries in row i of T are at least of size i, we can replace each $T_{i,j}$ by $T_{i,j} - i$ to obtain a new type of tableau.

Definition 4.2.1 A *reverse plane partition of shape* λ, T, is an array obtained by replacing the nodes of λ by nonnegative integers so that the rows and columns are weakly increasing. If the entries in of T sum to n, we say T is a *reverse plane partition of n*. Let

$rpp_\lambda(n) = $ the number of reverse plane partitions of n having shape λ ∎

The use of the term *reverse* for plane partitions where the parts *increase* is a historical accident stemming from the fact that ordinary partitions are usually written in weakly decreasing order.

From the definitions, we clearly have

$$\sum_{n\geq 0} ss_\lambda(n)x^n = x^{m(\lambda)} \sum_{n\geq 0} rpp_\lambda(n)x^n$$

where $m(\lambda) = \sum_{i\geq 1} i\lambda_i$. Thus it suffices to find the generating function for reverse plane partitions. Once again, the hooklengths come into play.

Theorem 4.2.2 *Fix a partition λ. Then*

$$\sum_{n\geq 0} rpp_\lambda(n)x^n = \prod_{(i,j)\in\lambda} \frac{1}{1 - x^{h_{i,j}}}$$

Proof. By the discussion after Proposition 4.1.4, the coefficient of x^n in this product counts partitions of n, where each part is of the form $h_{i,j}$ for some $(i,j) \in \lambda$. (Note that the part $h_{i,j}$ is associated with the node $(i,j) \in \lambda$, so parts $h_{i,j}$ and $h_{k,l}$ are considered different if $(i,j) \neq (k,l)$ even if $h_{i,j} = h_{k,l}$ as integers.) To show that this coefficient equals the number of reverse plane partitions T of n, it suffices to find a bijection

$$T \longleftrightarrow (h_{i_1,j_1}, h_{i_2,j_2}, \ldots)$$

which is weight preserving—i.e.,

$$\sum_{(i,j)\in\lambda} T_{i,j} = \sum_k h_{i_k,j_k}$$

$T \rightarrow (h_{i_1,j_1}, h_{i_2,j_2}, \ldots)$. Given T, we will produce a sequence of reverse plane partitions

$$T = T_0, T_1, T_2, \ldots, T_f = \text{ tableau of zeros}$$

where T_k will be obtained from T_{k-1} by subtracting one from all elements of a certain path of cells p_k in T_k. Since we will always have $|p_k| = h_{i_k,j_k}$ for some (i_k, j_k), this will ensure the weight-preserving condition.

Define the path $p = p_1$ in T inductively as follows.

HG1 Start p at (a, b), the rightmost highest cell of T containing a nonzero entry.

HG2 Continue by

$$\textbf{if } (i,j) \in p \textbf{ then} \begin{cases} (i, j-1) \in p & \text{if } T_{i,j-1} = T_{i,j} \\ (i+1, j) \in p & \text{otherwise} \end{cases}$$

In other words, move down unless forced to move left in order not to violate the weakly increasing condition along the rows (once the ones are subtracted).

HG3 Terminate p when the preceding induction rule fails. At this point we must be at the end of some column, say column c.

It is easy to see that after subtracting one from the elements in p, the array remains a reverse plane partition and the amount subtracted is $h_{a,c}$.

As an example, let

$$T = \begin{matrix} 1 & 2 & 2 & 2 \\ 3 & 3 & 3 & \\ 3 & & & \end{matrix}$$

Then $(a, b) = (1, 4)$ and the path p is indincated by the dotted cells in the following diagram:

After subtraction, we have

$$T_1 = \begin{matrix} 1 & 1 & 1 & 1 \\ 2 & 2 & 3 & \\ 2 & & & \end{matrix}$$

and $h_{i_1,j_1} = h_{1,1}$. To obtain the rest of the T_k, we iterate this process. The complete list for our example, together with the corresponding $h_{i,j}$, is

T_k:	1 2 2 2	1 1 1 1	0 0 0 0	0 0 0 0	0 0 0 0	0 0 0 0
	3 3 3	2 2 3	1 2 3	1 2 2	1 1 1	0 0 0
	3	2	1	1	1	0

h_{i_k,j_k}:		$h_{1,1}$	$h_{1,1}$	$h_{2,3}$	$h_{2,2}$	$h_{2,1}$

Thus $T \to (h_{1,1}, h_{1,1}, h_{2,3}, h_{2,2}, h_{2,1})$.

$\overline{(h_{i_1,j_1}, h_{i_2,j_2}, \ldots)} \to T$. Given a partition of hooklengths, we must rebuild the reverse plane partition. First, however, we must know in what order the hooklengths were removed.

Lemma 4.2.3 *In the decomposition of T into hooklengths, $h_{i,j}$ was removed before $h_{i',j'}$ if and only if*

$$i' > i, \quad or \quad i' = i \quad and \quad j' \le j \tag{4.3}$$

Proof. Since (4.3) is a total order on the nodes of the shape, we only need to prove the only-if direction. By transitivity, it suffices to consider the case where $h_{i',j'}$ is removed directly after $h_{i,j}$.

Let T and T' be the arrays from which $h_{i,j}$ and $h_{i',j'}$ were removed using paths p and p', respectively. By the choice of initial points and the fact that entries decrease in passing from T to T', we have $i' \ge i$.

If $i' > i$, we are done. Otherwise, $i' = i$ and p' starts in a column weakly to the left of p. We claim that in this case p' can never pass through a node strictly to the right of a node of p, forcing $j' \le j$. If not, then there is some $(s,t) \in p \cap p'$ such that $(s, t-1) \in p$ and $(s+1, t) \in p'$. But the fact that p moved left implies $T_{s,t} = T_{s,t-1}$. Since this equality continues to hold in T' after the ones have been subtracted, p' is forced to move left as well, a contradiction. ∎

Returning to the construction of T, if we are given a partition of hooklengths, then order them as in the lemma; suppose we get

$$(h_{i_1,j_1}, \ldots, h_{i_f,j_f})$$

We then construct a sequence of tableaux, starting with the all-zero array,

$$T_f, T_{f-1}, \ldots, T_0 = T$$

by adding back the h_{i_k,j_k} for $k = f, f-1, \ldots, 1$. To add $h_{a,c}$ to T, we construct a reverse path r along which to add ones.

GH1 Start r at the lowest node in column c.

GH2 Continue by

$$\textbf{if } (i,j) \in r \textbf{ then} \begin{cases} (i, j+1) \in r & \text{if } T_{i,j+1} = T_{i,j} \\ (i-1, j) \in r & \text{otherwise} \end{cases}$$

GH3 Terminate r when it passes through the rightmost node of the row a.

It is clear that this is a step-by-step inverse of the construction of the path p. However, it is not clear that r is well defined—i.e., that it must pass through the end of row a. Thus to finish the proof of Theorem 4.2.2, it suffices to prove a last lemma.

Lemma 4.2.4 *If r_k is the reverse path for h_{i_k,j_k}, then $(i_k, \lambda_{i_k}) \in r_k$.*

Proof. Use reverse induction on k. The result is obvious when $k = f$ by the first alternative in step GH2.

For $k < f$, let $r = r_k$ and $r' = r_{k+1}$. Similarly, define $T, T', h_{i,j}$, and $h_{i',j'}$. By our ordering of the hooklengths, $i \leq i'$. If $i < i'$, then the ith row of T consists solely of zeros, and we are done as in the base case.

If $i = i'$, then $j \geq j'$. Thus p starts weakly to the right of p'. By the same arguments as in Lemma 4.3, p stays to the right of p'. Since p' reaches the right end of row $i' = i$ by assumption, so must p. ∎

It is natural to ask if there is any relation between the hook formula and the hook generating function. In fact, the former is a corollary of the latter if we appeal to some general results of Stanley [Stn 71] about poset partitions.

Definition 4.2.5 Let $(A, \leq)$ be a partially ordered set. A *reverse A-partition* *of m* is an order-preserving map

$$\alpha : A \to \{0, 1, 2, \ldots\}$$

such that $\sum_{v \in A} \alpha(v) = m$. If $|A| = n$, then an order-preserving bijection

$$\beta : A \to \{1, 2, \ldots, n\}$$

is called a *natural labeling* of A. ∎

To see the connection with tableaux, partially order the cells of λ in the natural way,

$$(i, j) \leq (i', j') \iff i \leq i' \quad \text{and} \quad j \leq j'$$

Then a natural labeling of λ is just a standard λ-tableau, whereas a reverse λ-partition is a reverse plane partition of shape λ.

One of Stanley's theorems about poset partitions is

Theorem 4.2.6 ([Stn 71]) *Let A be a poset with $|A| = n$. Then the generating function for reverse A-partitions is*

$$\frac{P(x)}{(1 - x)(1 - x^2) \cdots (1 - x^n)}$$

where $P(x)$ is a polynomial such that $P(1)$ is the number of natural labelings of A. ∎

In the case where $A = \lambda$, we can compare this result with Theorem 4.2.2 and obtain

$$\frac{P(x)}{(1 - x)(1 - x^2) \cdots (1 - x^n)} = \prod_{(i,j) \in \lambda} \frac{1}{1 - x^{h_{i,j}}}$$

Thus

$$
\begin{aligned}
f^\lambda &= P(1) \\
&= \lim_{x \to 1} \frac{\prod_{k=1}^n (1 - x^k)}{\prod_{(i,j) \in \lambda} (1 - x^{h_{i,j}})} \\
&= \frac{n!}{\prod_{(i,j) \in \lambda} h_{i,j}}
\end{aligned}
$$

4.3 The Ring of Symmetric Functions

In order to keep track of more information with our generating functions, we can use more than one variable. The ring of symmetric functions then arises as a set of power series invariant under the action of all the symmetric groups.

Let $\mathbf{x} = \{x_1, x_2, x_3, \ldots\}$ be an infinite set of variables and consider the formal power series ring $\mathbf{C}[[\mathbf{x}]]$. The monomial $x_{i_1}^{\lambda_1} x_{i_2}^{\lambda_2} \cdots x_{i_l}^{\lambda_l}$ is said to have *degree* n if $n = \sum_i \lambda_i$. We also say that $f(\mathbf{x}) \in \mathbf{C}[[\mathbf{x}]]$ is *homogeneous of degree* n if every monomial in $f(\mathbf{x})$ has degree n.

For every n, there is a natural action of $\pi \in S_n$ on $f(\mathbf{x}) \in \mathbf{C}[[\mathbf{x}]]$, namely,

$$
\pi f(x_1, x_2, x_3, \ldots) = f(x_{\pi 1}, x_{\pi 2}, x_{\pi 3}, \ldots) \tag{4.4}
$$

where $\pi i = i$ for $i > n$. The simplest functions fixed by this action are gotten by symmetrizing a monomial.

Definition 4.3.1 Let $\lambda = (\lambda_1, \lambda_2, \ldots, \lambda_l)$ be a partition. The *monomial symmetric function corresponding to* λ is

$$
m_\lambda = m_\lambda(\mathbf{x}) = \sum x_{i_1}^{\lambda_1} x_{i_2}^{\lambda_2} \cdots x_{i_l}^{\lambda_l}
$$

where the sum is over all distinct monomials having exponents $\lambda_1, \ldots, \lambda_l$. ∎

For example

$$
m_{(2,1)} = x_1^2 x_2 + x_1 x_2^2 + x_1^2 x_3 + x_1 x_3^2 + x_2^2 x_3 + x_2 x_3^2 + \cdots
$$

Clearly, if $\lambda \vdash n$, then $m_\lambda(\mathbf{x})$ is homogeneous of degree n.

Now we can define the symmetric functions that interest us.

Definition 4.3.2 The *ring of symmetric functions* is

$$
\Lambda = \Lambda(\mathbf{x}) = \mathbf{C}[m_\lambda]
$$

—i.e., the vector space spanned by all the m_λ. ∎

Note that Λ is really a ring, not just a vector space, since it is closed under product. However, there are certain elements of $\mathbf{C}[[\mathbf{x}]]$ invariant under (4.4) that are not in Λ, such as $\prod_{i \geq 1} (1 + x_i)$, which cannot be written as a *finite* linear combination of m_λ.

We have the decomposition

$$\Lambda = \oplus_{n \geq 0} \Lambda^n$$

where Λ^n is the space spanned by all m_λ of degree n. In fact this is a *grading* of Λ since

$$f \in \Lambda^n \quad \text{and} \quad g \in \Lambda^m \quad \text{implies} \quad fg \in \Lambda^{n+m} \tag{4.5}$$

Since the m_λ are independent, we have the following result.

Proposition 4.3.3 *The space Λ^n has basis*

$$\{m_\lambda \mid \lambda \vdash n\}$$

and so has dimension $p(n)$, the number of partitions of n. ∎

There are several other bases for Λ^n that are of interest. To construct them, we need the following families of symmetric functions.

Definition 4.3.4 The nth *power sum symmetric function* is

$$p_n = m_{(n)} = \sum_{i \geq 1} x_i^n$$

The nth *elementary symmetric function* is

$$e_n = m_{(1^n)} = \sum_{i_1 < \cdots < i_n} x_{i_1} \cdots x_{i_n}$$

The nth *complete homogeneous symmetric function* is

$$h_n = \sum_{\lambda \vdash n} m_\lambda = \sum_{i_1 \leq \cdots \leq i_n} x_{i_1} \cdots x_{i_n} \quad \blacksquare$$

As examples, when $n = 3$

$$
\begin{aligned}
p_3 &= x_1^3 + x_2^3 + x_3^3 + \cdots \\
e_3 &= x_1 x_2 x_3 + x_1 x_2 x_4 + x_1 x_3 x_4 + x_2 x_3 x_4 + \cdots \\
h_3 &= x_1^3 + x_2^3 + \cdots + x_1^2 x_2 + x_1 x_2^2 + \cdots + x_1 x_2 x_3 + x_1 x_2 x_4 + \cdots
\end{aligned}
$$

The elementary function e_n is just the sum of all square-free monomials of degree n. As such, it can be considered as a weight generating function for partitions with n distinct parts. Specifically, let

$$S = \{\lambda \mid l(\lambda) = n\}$$

where $l(\lambda)$ is the number of parts of λ, known as its *length*. If $\lambda = (\lambda_1 > \lambda_2 > \cdots > \lambda_n)$, we use the weight

$$\text{wt } \lambda = x_{\lambda_1} x_{\lambda_2} \cdots x_{\lambda_n}$$

which yields

$$e_n(\mathbf{x}) = f_S(\mathbf{x})$$

Similarly, h_n is the sum of all monomials of degree n and is the weight generating function for all partitions with n parts. What if we want to count partitions with any number of parts?

Proposition 4.3.5 *We have the following generating functions*

$$E(t) \overset{\text{def}}{=} \sum_{n \geq 0} e_n(\mathbf{x}) t^n = \prod_{i \geq 1} (1 + x_i t)$$

$$H(t) \overset{\text{def}}{=} \sum_{n \geq 0} h_n(\mathbf{x}) t^n = \prod_{i \geq 1} \frac{1}{(1 - x_i t)}$$

Proof. Work in the ring $\mathbf{C}[[\mathbf{x}, t]]$. For the elementary symmetric functions, consider the set $S = \{\lambda \mid \lambda \text{ with distinct parts}\}$ with weight

$$\text{wt}' \, \lambda = t^{l(\lambda)} \, \text{wt} \, \lambda$$

where wt is as before. Then

$$
\begin{aligned}
f_S(\mathbf{x}, t) &= \sum_{\lambda \in S} \text{wt}' \, \lambda \\
&= \sum_{n \geq 0} \sum_{l(\lambda) = n} t^n \, \text{wt} \, \lambda \\
&= \sum_{n \geq 0} e_n(\mathbf{x}) t^n
\end{aligned}
$$

To obtain the product, write

$$S = (\{1^0\} \uplus \{1^1\}) \times (\{2^0\} \uplus \{2^1\}) \times (\{3^0\} \uplus \{3^1\}) \times \cdots$$

so that

$$f_S(\mathbf{x}, t) = (1 + x_1 t)(1 + x_2 t)(1 + x_3 t) \cdots$$

The proof for the complete symmetric functions is analogous. ∎

While we are computing generating functions, we might as well give one for the power sums. Actually, it is easier to produce one for $p_n(\mathbf{x})/n$.

Proposition 4.3.6 *We have the following generating function*

$$\sum_{n \geq 1} p_n(\mathbf{x}) \frac{t^n}{n} = \ln \prod_{i \geq 1} \frac{1}{(1 - x_i t)}$$

Proof. Using the Taylor expansion of $\ln \frac{1}{1-x}$:

$$
\begin{aligned}
\ln \prod_{i \geq 1} \frac{1}{(1 - x_i t)} &= \sum_{i \geq 1} \ln \frac{1}{(1 - x_i t)} \\
&= \sum_{i \geq 1} \sum_{n \geq 1} \frac{(x_i t)^n}{n} \\
&= \sum_{n \geq 1} \frac{t^n}{n} \sum_{i \geq 1} x_i^n \\
&= \sum_{n \geq 1} p_n(\mathbf{x}) \frac{t^n}{n} \quad \blacksquare
\end{aligned}
$$

In order to have enough elements for a basis of Λ^n, we must have one function for each partition of n according to Proposition 4.3.3. To extend Definition 4.3.4 to $\lambda = (\lambda_1, \lambda_2, \ldots, \lambda_l)$, let

$$f_\lambda = f_{\lambda_1} f_{\lambda_2} \cdots f_{\lambda_l}$$

where $f = p, e$, or h. We say these functions are *multiplicative*. To illustrate, if $\lambda = (2, 1)$, then

$$p_{(2,1)} = (x_1^2 + x_2^2 + x_3^2 + \cdots)(x_1 + x_2 + x_3 + \cdots)$$

Theorem 4.3.7 *The following are bases for* Λ^n.

1. $\{p_\lambda \mid \lambda \vdash n\}$

2. $\{e_\lambda \mid \lambda \vdash n\}$

3. $\{h_\lambda \mid \lambda \vdash n\}$

Proof.

1. Let $C = (c_{\lambda\mu})$ be the matrix expressing the p_λ in terms of the basis m_μ. If we can find an ordering of partitions so that C is triangular with nonzero entries down the diagonal, then C^{-1} exists, and the p_λ are also a basis. It turns out that lexicographic order will work. In fact, we claim that

$$p_\lambda = c_{\lambda\lambda} m_\lambda + \sum_{\mu \rhd \lambda} c_{\lambda\mu} m_\mu \qquad (4.6)$$

where $c_{\lambda\lambda} \neq 0$. (This is actually stronger than our claim about C by Proposition 2.2.6.) But if $\mathbf{x}_1^{\mu_1} \mathbf{x}_2^{\mu_2} \cdots \mathbf{x}_m^{\mu_m}$ appears in

$$p_\lambda = (x_1^{\lambda_1} + x_2^{\lambda_1} + \cdots)(x_1^{\lambda_2} + x_2^{\lambda_2} + \cdots) \cdots$$

then each μ_i must be a sum of λ_j's. Since adding together parts of a partition makes it become larger in dominance order, m_λ must be the smallest term that occurs.

2. In a similar manner we can show that there exist scalars $d_{\lambda\mu}$ such that

$$e_{\lambda'} = m_\lambda + \sum_{\mu \lhd \lambda} d_{\lambda\mu} m_\mu$$

where λ' is the conjugate of λ.

3. Since there are $p(n) = \dim \Lambda^n$ functions h_λ, it suffices to show that they generate the basis e_μ. Since both sets of functions are multiplicative, we may simply demonstrate that every e_n is a polynomial in the h_k. From the products in Proposition 4.3.5, we see that

$$H(t)E(-t) = 1$$

Substituting in the summations for H and E and picking out the coefficient of t^n on both sides yields

$$\sum_{r=0}^{n}(-1)^r h_{n-r} e_r = 0$$

for $n \geq 1$. So

$$e_n = h_1 e_{n-1} - h_2 e_{n-2} + \cdots$$

which is a polynomial in the h's by induction on n. ∎

Part 2 of this theorem is often called the fundamental theorem of symmetric functions and stated as: every symmetric function is a polynomial in the elementary functions e_n.

4.4 Schur Functions

There is a fifth basis for Λ^n that is very important, the Schur functions. As we will see, they are also intimately connected with the irreducible representations of $\mathcal{S}_n$ and tableaux. In fact, they are so protean that there are many different ways to define them. In this section, we take the combinatorial approach.

Given any composition $\mu = (\mu_1, \mu_2, \dots, \mu_l)$, there is a corresponding monomial weight in $\mathbf{C}[[\mathbf{x}]]$:

$$\mathbf{x}^\mu \stackrel{\text{def}}{=} x_1^{\mu_1} x_2^{\mu_2} \cdots x_m^{\mu_l} \tag{4.7}$$

Now consider any generalized tableau T of shape λ. It also has a weight—namely,

$$\mathbf{x}^T \stackrel{\text{def}}{=} \prod_{(i,j)\in\lambda} x_{T_{i,j}} = \mathbf{x}^\mu \tag{4.8}$$

where μ is the content of T. For example, if

$$T = \begin{array}{ccc} 4 & 1 & 4 \\ 1 & 3 \end{array}$$

then

$$\mathbf{x}^T = x_1^2 x_3 x_4^2$$

Definition 4.4.1 Given a partition λ, the associated *Schur function* is

$$s_\lambda(\mathbf{x}) = \sum_T \mathbf{x}^T$$

where the sum is over all semistandard λ-tableaux T.

By way of illustration, if $\lambda = (2,1)$, then some of the possible tableaux are

$$\begin{array}{llllllllll}
1\,1, & 1\,2, & 1\,1, & 1\,3, & \dots & 1\,2, & 1\,3, & 1\,2, & 1\,4, & \dots \\
2 & 2 & 3 & 3 & & 3 & 2 & 4 & 2 &
\end{array}$$

so

$$s_{(2,1)}(\mathbf{x}) = x_1^2 x_2 + x_1 x_2^2 + x_1^2 x_3 + x_1 x_3^2 + \cdots + 2x_1 x_2 x_3 + 2x_1 x_2 x_4 + \cdots$$

Note that if $\lambda = (n)$, then a one-rowed tableau is just a weakly increasing sequence of n positive integers, i.e., a partition with n parts (written backward), so

$$s_{(n)}(\mathbf{x}) = h_n(\mathbf{x}) \tag{4.9}$$

If we have only one column, then the entries must increase from top to bottom, so the partition must have distinct parts and thus

$$s_{(1^n)} = e_n(\mathbf{x}) \tag{4.10}$$

Finally, if $\lambda \vdash n$ is arbitrary, then

$$[x_1 x_2 \cdots x_n] s_\lambda(\mathbf{x}) = f^\lambda$$

since pulling out this coefficient merely looks at standard tableaux.

Before we can show that the s_λ are a basis for Λ^n, we must verify that they are indeed symmetric functions. We give two proofs of this fact, one based on our results from representation theory and one combinatorial (the latter being due to Knuth [Knu 70]).

Proposition 4.4.2 *The function $s_\lambda(\mathbf{x})$ is symmetric.*

Proof 1. By definition of the Schur functions and Kostka numbers,

$$s_\lambda = \sum_\mu K_{\lambda\mu} \mathbf{x}^\mu \tag{4.11}$$

where the sum is over all compositions μ of n. Thus it is enough to show that

$$K_{\lambda\mu} = K_{\lambda\tilde{\mu}} \tag{4.12}$$

for any rearrangement $\tilde{\mu}$ of μ. But in this case M^μ and $M^{\tilde{\mu}}$ are isomorphic modules. Thus they have the same decomposition into irreducibles, and (4.12) follows from Young's rule (Theorem 2.11.2).

Proof 2. It suffices to show that

$$(i, i+1)s_\lambda(\mathbf{x}) = s_\lambda(\mathbf{x})$$

for each adjacent transposition. To this end, we describe an involution on semistandard λ-tableaux

$$T \longrightarrow T'$$

such that the numbers of i's and $i+1$s are exchanged when passing from T
to T' (with all other multiplicities staying the same).

Given T, each column contains either an $i, i+1$ pair; exactly one of $i, i+1$;
or neither. Call the pairs *fixed* and all other occurrences of i or $i+1$ *free*. In
each row switch the number of free i's and $i+1$s—i.e., if the the row consists
of k free i's followed by l free $i+1$s replace them by l free i's followed by k
free $i+1$s. To illustrate, if $i = 2$ and

$$T = \begin{matrix} 1 & 1 & 1 & 1 & 2 & 2 & 2 & 2 & 2 & 3 \\ 2 & 2 & 3 & 3 & 3 & 3 \\ 3 \end{matrix}$$

then the twos and threes in columns 2 through 4 and 7 through 10 are free.
So

$$T' = \begin{matrix} 1 & 1 & 1 & 1 & 2 & 2 & 2 & 3 & 3 & 3 \\ 2 & 2 & 2 & 3 & 3 & 3 \\ 3 \end{matrix}$$

The new tableau T' is still semistandard by the definition of free. Since the
fixed i's and $i+1$s come in pairs, this map has the desired exchange property.
It is also clearly an involution. ∎

Using the ideas in the proof of Theorem 4.3.7, part 1, the following result
guarantees that the s_λ are a basis.

Proposition 4.4.3 *We have*

$$s_\lambda = \sum_{\mu \trianglelefteq \lambda} K_{\lambda\mu} m_\mu$$

where the sum is over partitions μ (rather than compositions) and $K_{\lambda\lambda} = 1$.

Proof. By equation (4.11) and the symmetry of the Schur functions, we have

$$s_\lambda = \sum_\mu K_{\lambda\mu} m_\mu$$

where the sum is over all *partitions* μ. We can prove that

$$K_{\lambda\mu} = \begin{cases} 0 & \text{if } \lambda \ntrianglerighteq \mu \\ 1 & \text{if } \lambda = \mu \end{cases}$$

in two different ways.

One is to appeal again to Young's rule and Corollary 2.4.7. The other is
combinatorial. If $K_{\lambda\mu} \neq 0$, then consider a λ-tableau T of content μ. Since
T is column-strict, all occurrences of the numbers $1, 2, \ldots, i$ are in rows 1
through i. This implies that for all i,

$$\mu_1 + \mu_2 + \cdots + \mu_i \leq \lambda_1 + \lambda_2 + \cdots + \lambda_i$$

—i.e., $\mu \trianglelefteq \lambda$. Furthermore, if $\lambda = \mu$, then by the same reasoning there is only
one tableau of shape and content λ—namely, the one where row i contains
all occurrences of i. (Some authors call this tableau *superstandard*.) ∎

Corollary 4.4.4 *The set $\{s_\lambda \mid \lambda \vdash n\}$ is a basis for Λ^n.* ∎

4.5   The Jacobi-Trudi Determinants

The determinantal formula (Theroem 3.2.1) calculated the number of standard tableau, f^λ. Analogously, the Jacobi-Trudi determinants provide another expression for s_λ in terms of elementary and complete symmetric functions. Jacobi [Jac 41] was the first to obtain this result, and his student Trudi [Tru 64] subsequently simplified it.

We have already seen the special 1×1 case of these determinants in equations (4.9) and (4.10). The general result is as follows. Any symmetric function with a negative subscript is defined to be zero.

Theorem 4.5.1 (Jacobi-Trudi Determinants) *Let* $\lambda = (\lambda_1, \lambda_2, \ldots, \lambda_l)$. *We have*

$$s_\lambda = |h_{\lambda_i - i + j}|$$

and

$$s_{\lambda'} = |e_{\lambda_i - i + j}|$$

where λ' is the conjugate of λ and both determinants are $l \times l$.

Proof. We prove this theorem using a method of Gessel [Ges um] and Gessel-Viennot [G-V 85, G-V ip] that permits us to view both tableaux and determinants as lattice paths. Consider the plane $\mathbf{Z} \times \mathbf{Z}$ of integer lattice points. We consider (possibly infinite) paths in this plane

$$p = s_1, s_2, s_3, \ldots$$

where each step s_i is of unit length north (N) or east (E). Such a path is shown in the following figure.

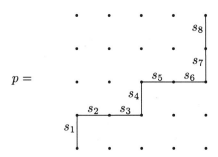

Label the eastern steps of p using one of two labelings. The *e-labeling* assigns to each eastern s_i the label

$$L(s_i) = i$$

The *h-labeling* gives s_i the label

$$\check{L}(s_i) = \text{(the number of northern } s_j \text{ preceding } s_i) + 1$$

Intuitively, in the h-labeling all the eastern steps on the line through the origin of p are labeled 1, all those on the line one unit above are labeled 2, and so on. Labeling our example path with each of the two possibilities yields the next pair of diagrams.

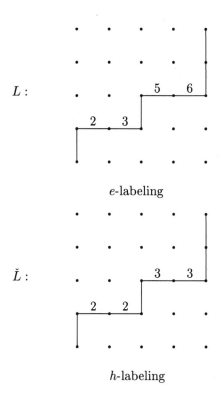

e-labeling

h-labeling

It is convenient to extend $\mathbf{Z} \times \mathbf{Z}$ by the addition of some points at infinity. Specifically, for each $x \in \mathbf{Z}$, add a point (x, ∞) above every point on the vertical line with coordinate x. We assume that a path can reach (x, ∞) only by ending with an infinite number of consecutive northern steps along this line. If p starts at a vertex u and ends at a vertex v (which may be a point at infinity), then we write $u \xrightarrow{p} v$.

There are two weightings of paths corresponding to the two labelings. If p has only a finite number of eastern steps, define

$$\mathbf{x}^p = \prod_{s_i \in p} x_{L(s_i)}$$

and

$$\check{\mathbf{x}}^p = \prod_{s_i \in p} x_{\check{L}(s_i)}$$

where each product is taken over the eastern s_i in p. Note that $\mathbf{x}^p$ is always square-free and $\check{\mathbf{x}}^p$ can be any monomial. So we have

$$e_n(\mathbf{x}) = \sum_p \mathbf{x}^p$$

and

$$h_n(\mathbf{x}) = \sum_p \check{\mathbf{x}}^p$$

where both sums are over all paths $(a, b) \xrightarrow{p} (a + n, \infty)$ for any fixed initial vertex (a, b).

Just as all paths between one pair of points describes a lone elementary or complete symmetric function, all l-tuples of paths between l pairs of points will be used to model the l-fold products contained in the Jacobi-Trudi determinants. Let $u_1, u_2, \ldots, u_l$ and $v_1, v_2, \ldots, v_l$ be fixed sets of initial and final vertices. Consider a family of paths $\mathcal{P} = (p_1, p_2, \ldots, p_l)$, where, for each i,

$$u_i \xrightarrow{p_i} v_{\pi i}$$

for some $\pi \in S_l$. Give weight to a family by letting

$$\mathbf{x}^{\mathcal{P}} = \prod_i \mathbf{x}^{p_i}$$

with a similar definition for $\check{\mathbf{x}}^{\mathcal{P}}$. Also define the *sign of* $\mathcal{P}$ to be

$$(-1)^{\mathcal{P}} = \operatorname{sgn} \pi$$

For example, the family

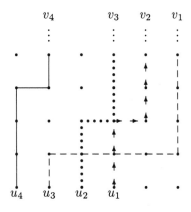

has

$$\check{\mathbf{x}}^{\mathcal{P}} = x_2^4 x_3^2 x_4$$

and sign

$$(-1)^{\mathcal{P}} = \operatorname{sgn}(1, 2, 3)(4) = +1$$

We now concentrate on proving

$$s_\lambda = |h_{\lambda_i - i + j}| \tag{4.13}$$

the details for the elementary determinant being similar and left to the reader. Given λ, pick initial points

$$u_i = (1 - i, 0) \tag{4.14}$$

and the final ones

$$v_i = (\lambda_i - i + 1, \infty) \tag{4.15}$$

The preceding 4-tuple is an example for $\lambda = (2, 2, 2, 1)$. From the choice of vertices,

$$h_{\lambda_i - i + j} = \sum_{u_j \xrightarrow{P} v_i} \check{x}^P$$

Thus the set of l-tuples $\mathcal{P}$ with permutation π corresponds to the term in the determinant obtained from the entries with coordinates

$$(\pi 1, 1), (\pi 2, 2), \ldots, (\pi l, l)$$

So

$$|h_{\lambda_i - i + j}(\mathbf{x})| = \sum_{\mathcal{P}} (-1)^P \check{x}^P \tag{4.16}$$

where the sum is over all families of paths with initial points and final points given by the u_j and v_i.

Next we show that all the terms on the right side of equation (4.16) cancel in pairs except for those corresponding to l-tuples of nonintersecting paths. To do this, we need a weight-preserving involution

$$\mathcal{P} \xleftrightarrow{\iota} \mathcal{P}'$$

such that if $\mathcal{P} = \mathcal{P}'$ (corresponding to the 1-cycles of ι), then $\mathcal{P}$ is nonintersecting and if $\mathcal{P} \neq \mathcal{P}'$ (corresponding to the 2-cycles), then $(-1)^P = -(-1)^{P'}$. The basic idea is that if $\mathcal{P}$ contains some intersections, then we will find two uniquely defined intersecting paths and switch their final portions.

Definition 4.5.2 Given $\mathcal{P}$; define $\iota\mathcal{P} = \mathcal{P}'$, where

1. if $p_i \cap p_j = \emptyset$ for all i, j, then $\mathcal{P} = \mathcal{P}'$,

2. otherwise, find the smallest index i such that the path p_i intersects some other path. Let v_0 be the first (SW-most) intersection on p_i and let p_j be the other path through v_0. (If there is more than one, choose p_j so that k is minimal.) Now define

$$\mathcal{P}' = \mathcal{P} \qquad \text{with } p_i, p_j \text{ replaced by } p_i', p_j'$$

where

$$p_i' = u_i \xrightarrow{p_i} v_0 \xrightarrow{p_j} v_{\pi j} \quad \text{and} \quad p_j' = u_j \xrightarrow{p_j} v_0 \xrightarrow{p_i} v_{\pi i} \blacksquare$$

By way of illustration, we can apply this map to our 4-tuple.

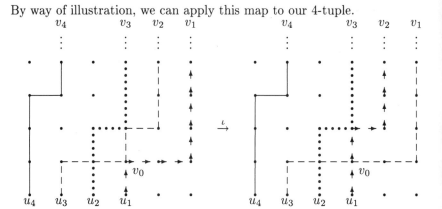

Our choice of p_i, p_j, and v_0 make it possible to reconstruct them after applying ι, so the map is invertible with itself as inverse. The nonintersecting families are fixed by definition. For the intersecting l-tuples, the sign clearly changes when passing from $\mathcal{P}$ to $\mathcal{P}'$. Weight is preserved due to the fact that all paths start on the same horizontal axis. Thus the set of labels is unchanged by ι.

Because of the cancellation in (4.16), the only thing that remains is to show that

$$s_\lambda(\mathbf{x}) = \sum_{\mathcal{P}} (-1)^{\mathcal{P}} \check{\mathbf{x}}^{\mathcal{P}}$$

where the sum is now over all l-tuples of nonintersecting paths. But by our choice of initial and final vertices, $\mathcal{P} = (p_1, p_2, \ldots, p_l)$ is nonintersecting only if it corresponds to the identity permutation. Thus $(-1)^{\mathcal{P}} = +1$ and $u_i \xrightarrow{p_i} v_i$ for all i. There is a simple bijection between such families and semistandard tableaux. Given $\mathcal{P}$, merely use the h-labels of the ith path, listed in increasing order, for the ith row of a tableau T. For example,

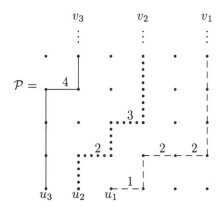

becomes the tableau

$$T = \begin{array}{ccc} 1 & 2 & 2 \\ 2 & 3 & \\ 4 & & \end{array}$$

In view of equations (4.14) and (4.15), T has shape λ. By definition the rows of T weakly increase. Finally, the columns must strictly increase because the nonintersecting condition and choice of initial vertices force the jth eastern step of p_{i+1} to be higher than the corresponding step on p_i. Construction of the inverse map is an easy exercise. This completes the proof of (4.13). ∎

4.6 Other Definitions of the Schur Function

We would be remiss if we did not give the definition that Schur originally used [Scu 01] (as a quotient of alternants) for the functions that bear his name. This definition actually goes back to Jacobi [Jac 41], but Schur was the first to notice the connection with irreducible characters of $\mathcal{S}_n$, which is our third way of defining s_λ.

First, we have to restrict Λ to l variables, where $l = l(\lambda)$. (Actually any $l \geq l(\lambda)$ will do for s_λ.) Specifically, let

$$\Lambda_l = \{f(x_1,\ldots,x_l,0,0,\ldots) \mid f(\mathbf{x}) \in \Lambda\}$$

We use the abbreviation $f(x_1, x_2, \ldots, x_l)$ for a typical element of Λ_l. For example,

$$p_3(x_1, x_2, x_3, x_4) = x_1^3 + x_2^3 + x_3^3 + x_4^3$$

Thus Λ_l is just the set of functions in the polynomial ring $\mathbf{C}[x_1, x_2, \ldots, x_l]$ that are fixed by the action (4.4) of $\mathcal{S}_l$.

Similarly we can consider the *skew-symmetric functions* in $\mathbf{C}[x_1, x_2, \ldots, x_l]$, which are those satisfying

$$\pi f = (\operatorname{sgn} \pi) f$$

for all $\pi \in \mathcal{S}_l$. Just as we can obtain symmetric functions by symmetrizing a monomial, we can obtain skew-symmetric ones by skew-symmetrization. Let $\mu = (\mu_1, \mu_2, \ldots, \mu_l)$ be any composition with monomial $\mathbf{x}^\mu$ given by (4.7). Define the corresponding *alternant* by

$$a_\mu(x_1, \ldots, x_l) = \sum_{\pi \in \mathcal{S}_l} (\operatorname{sgn} \pi) \pi \mathbf{x}^\mu$$

It is easy to verify that a_μ is skew-symmetric. From the definition of a determinant, we can also write

$$a_\mu = \left| x_i^{\mu_j} \right|_{1 \leq i,j \leq l} \tag{4.17}$$

For example,

$$a_{(4,2,1)}(x_1, x_2, x_3) = x_1^4 x_2^2 x_3 + x_1^2 x_2 x_3^4 + x_1 x_2^4 x_3^2 - x_1^4 x_2 x_3^2 - x_1^2 x_2^4 x_3 - x_1 x_2^2 x_3^4$$

$$= \begin{vmatrix} x_1^4 & x_1^2 & x_1 \\ x_2^4 & x_2^2 & x_2 \\ x_3^4 & x_3^2 & x_3 \end{vmatrix}$$

The most famous example of an alternant is when we let the composition be

$$\delta \stackrel{\text{def}}{=} (l-1, l-2, \ldots, 1, 0)$$

In this case

$$a_\delta = \left| x_i^{l-j} \right|$$

is the Vandermonde determinant. It is well-known that we have the factorization

$$a_\delta = \prod_{1 \le i < j \le l} (x_i - x_j) \tag{4.18}$$

If λ is any partition of length l with two equal parts, then $a_\lambda = 0$, since determinant (4.17) has two equal columns. So it does no harm to restrict our attention to partitions with distinct parts. These all have the form $\lambda + \delta$, where λ is an arbitrary partition and addition of integer vectors is done componentwise. Furthermore, if we set $x_i = x_j$ in any alternant, then we have two equal rows. So $a_{\lambda+\delta}$ is divisible by all the terms in product (4.18), which makes $a_{\lambda+\delta}/a_\delta$ a polynomial. In fact, it must be symmetric, being the quotient of skew-symmetric polynomials. The amazing thing, at least from our point of view, is that this is actually the symmetric function s_λ. To demonstrate this we follow the proof in Macdonald [Mac 79, page 25]. For $1 \le j \le l$, let $e_n^{(j)}$ denote the elementary symmetric function in the variables $x_1, \ldots, x_{j-1}, x_{j+1}, \ldots, x_l$ (x_j omitted). We have the following lemma.

Lemma 4.6.1 Let $\mu = (\mu_1, \mu_2, \ldots, \mu_l)$ be any composition. Consider the $l \times l$ matrices

$$A_\mu = (x_j^{\mu_i}), H_\mu = (h_{\mu_i - l + j}) \quad and \quad E = ((-1)^{l-i} e_{l-i}^{(j)})$$

Then

$$A_\mu = H_\mu E$$

Proof. Consider the generating function for the $e_n^{(j)}$,

$$E^{(j)}(t) \stackrel{\text{def}}{=} \sum_{n=0}^{l-1} e_n^{(j)} t^n = \prod_{i \ne j} (1 + x_i t)$$

We can now mimic the proof of Theorem 4.3.7, part 3. Since

$$H(t) E^{(j)}(-t) = \frac{1}{1 - x_j t}$$

we can extract the coefficient of t^{μ_i} on both sides. This yields

$$\sum_{k=1}^{l} h_{\mu_i - l + k} \cdot (-1)^{l-k} e_{l-k}^{(j)} = x_j^{\mu_i}$$

which is equivalent to what we wished to prove. ∎

Corollary 4.6.2 *Let λ have length l. Then*

$$s_\lambda = \frac{a_{\lambda + \delta}}{a_\delta}$$

where all functions are polynomials in $x_1, \ldots, x_l$.

Proof. Taking determinants in the lemma,

$$|A_\mu| = |H_\mu| \cdot |E| \tag{4.19}$$

where $|A_\mu| = a_\mu$. First of all, let $\mu = \delta$. In this case $H_\delta = (h_{i-j})$, which is upper unitriangular and thus has determinant 1. Plugging this into (4.19) gives $|E| = a_\delta$.

Now letting $\mu = \lambda + \delta$ in the same equation:

$$\frac{a_{\lambda+\delta}}{a_\delta} = |H_{\lambda+\delta}| = |h_{\lambda_i - i + j}|$$

Hence we are done, by the Jacobi-Trudi theorem. ∎

Our last description of the s_λ will involve the characters of $\mathcal{S}_n$. To see the connection, let us reexamine the change-of-basis matrix between the monomial and power sum symmetric functions introduced in (4.6). Let us compute a small example. When $n = 3$, we obtain

$$
\begin{array}{lclcl}
p_{(3)} & = & x_1^3 + x_2^3 + \cdots & = & m_{(3)} \\
p_{(2,1)} & = & (x_1^2 + x_2^2 + \cdots)(x_1 + x_2 + \cdots) & = & m_{(3)} + m_{(2,1)} \\
p_{(1^3)} & = & (x_1 + x_2 + \cdots)^3 & = & m_{(3)} + 3m_{(2,1)} + 6m_{(1^3)}
\end{array}
$$

Comparing this with the character values ϕ^μ of the permutation modules M^μ in Example 2.1.9, the reader will be lead to suspect the following theorem.

Theorem 4.6.3 *Let ϕ_λ^μ be character of M^μ evaluated on the class corresponding to λ. Then*

$$p_\lambda = \sum_{\mu \trianglerighteq \lambda} \phi_\lambda^\mu m_\mu \tag{4.20}$$

Proof. Let $\lambda = (\lambda_1, \lambda_2, \ldots, \lambda_l)$. Then we can write equation (4.6) as

$$\prod_i (x_1^{\lambda_i} + x_2^{\lambda_i} + \cdots) = \sum_\mu c_{\lambda\mu} m_\mu$$

Pick out the coefficient of $\mathbf{x}^\mu$ on both sides, where $\mu = (\mu_1, \mu_2, \ldots, \mu_m)$. On the right, it is $c_{\lambda\mu}$. On the left, it is the number of ways to distribute the parts of λ into subpartitions $\lambda^1, \ldots, \lambda^m$ such that

$$\biguplus_i \lambda^i = \lambda \quad \text{and} \quad \lambda^i \vdash \mu_i \quad \text{for all } i \tag{4.21}$$

where equal parts of λ are distinguished in order to be considered different in the disjoint union.

Now consider $\phi^\mu_\lambda = \phi^\mu(\pi)$, where $\pi \in \mathcal{S}_n$ is an element of cycle type λ. By definition, this character value is the number of fixed-points of the action of π on all standard tabloids t of shape μ. But t is fixed if and only if each cycle of π lies in a single row of t. Thus we must distribute the cycles of length λ_i among the rows of length μ_j subject to exactly the same restrictions as in (4.21). It follows that $c_{\lambda\mu} = \phi^\mu_\lambda$, as desired. ∎

Equation (4.20) shows that p_λ is the generating function for the character values on a fixed conjugacy class K_λ as the module in question varies over all M^μ. What we would like is a generating function for the fixed character χ^λ of the irreducible Specht module S^λ as K_μ varies over all conjugacy classes. To isolate χ^λ in (4.20), we use the inner product

$$\begin{aligned}
\langle \phi, \chi \rangle &= \frac{1}{n!} \sum_{\pi \in \mathcal{S}_n} \phi(\pi)\chi(\pi) \\
&= \text{the multiplicity of } \chi \text{ in } \phi
\end{aligned}$$

as long as χ is irreducible (Corollary 1.9.4, part 2).

If $\pi \in \mathcal{S}_n$ has type λ, then it will be convenient to define the corresponding power sum symmetric function $p_\pi = p_\lambda$; similar definitions can be made for the other bases. For example, if $\pi = (1,3,4)(2,5)$, then $p_\pi = p_{(3,2)}$. From the previous theorem,

$$p_\pi = \sum_\mu \phi^\mu(\pi)m_\mu$$

To introduce the inner product, we multiply by $\chi^\lambda(\pi)/n!$ and sum to get

$$\begin{aligned}
\frac{1}{n!} \sum_{\pi \in \mathcal{S}_n} p_\pi \chi^\lambda(\pi) &= \frac{1}{n!} \sum_{\pi \in \mathcal{S}_n} \left(\sum_\mu \phi^\mu(\pi)m_\mu \right) \chi^\lambda(\pi) \\
&= \sum_\mu m_\mu \left(\frac{1}{n!} \sum_\pi \phi^\mu(\pi)\chi^\lambda(\pi) \right) \\
&= \sum_\mu K_{\lambda\mu}m_\mu \qquad \text{(Young's rule)} \\
&= s_\lambda \qquad\qquad\qquad \text{(Proposition 4.4.3)}
\end{aligned}$$

We have proved the following theorem of Frobenius.

Theorem 4.6.4 *If $\lambda \vdash n$, then*

$$s_\lambda = \frac{1}{n!} \sum_{\pi \in \mathcal{S}_n} \chi^\lambda(\pi)p_\pi \quad \blacksquare \tag{4.22}$$

Note that this is a slightly different type of generating function from those discussed previously, using the power sum basis and averaging over $\mathcal{S}_n$. Such functions occur in other areas of combinatorics, notably Pólya theory, where they are related to cycle index polynomials of groups [PTW 83, pages 55–85].

There are several other ways to write equation (4.22). Since χ^λ is a class function, we can collect terms and obtain

$$s_\lambda = \frac{1}{n!} \sum_\mu k_\mu \chi^\lambda_\mu p_\mu$$

where $k_\mu = |K_\mu|$ and χ^λ_μ is the value of χ^λ on K_μ. Alternatively, we can use formula (1.2) to express things in terms of the size of centralizers:

$$s_\lambda = \sum_\mu \frac{1}{z_\mu} \chi^\lambda_\mu p_\mu \tag{4.23}$$

4.7 The Characteristic Map

Let $R^n = R(\mathcal{S}_n)$ be the space of class functions on $\mathcal{S}_n$. Then there is an intimate connection between R^n and Λ^n that we will explore in this section.

First of all, $\dim R^n = \dim \Lambda^n = p(n)$ (the number of partitions of n), so these two are isomorphic as vector spaces. We also have an inner product on R^n for which the irreducible characters on $\mathcal{S}_n$ form an orthonormal basis (Theorem 1.9.3). Motivated by equation (4.22), define an inner product on Λ^n by

$$\langle s_\lambda, s_\mu \rangle = \delta_{\lambda\mu}$$

and sesquilinear extension (linear in the first variable and conjugate linear in the second).

We now define a map to preserve these inner products.

Definition 4.7.1 The *characteristic map* is $\mathrm{ch}^n : R^n \to \Lambda^n$ defined by

$$\mathrm{ch}^n(\chi) = \sum_{\mu \vdash n} z_\mu^{-1} \chi_\mu p_\mu$$

where χ_μ is the value of χ on the class μ.

It is easy to verify that ch^n is linear. Furthermore, if we apply ch^n to the irreducible characters, then by equation (4.23)

$$\mathrm{ch}^n(\chi^\lambda) = s_\lambda$$

Since ch^n takes one orthonormal basis to another, we immediately have the following.

Proposition 4.7.2 *The map* ch^n *is an isometry between* R^n *and* Λ^n. ∎

Now consider $R = \oplus_n R^n$, which is isomorphic to $\Lambda = \oplus_n \Lambda^n$ via the characteristic map $\mathrm{ch} = \oplus_n \mathrm{ch}^n$. But Λ also has the structure of a graded algebra—i.e., a ring product satisfying (4.5). How can we construct a corresponding product in R^n? If χ and ψ are characters of $\mathcal{S}_n$ and $\mathcal{S}_m$, respectively, we want to produce a character of $\mathcal{S}_{n+m}$. But the tensor product $\chi \otimes \psi$ gives us a character of $\mathcal{S}_n \times \mathcal{S}_m$ and induction gets us into the group we want. Therefore, define a product on R by bilinearly extending

$$\chi \cdot \psi = (\chi \otimes \psi)\!\uparrow^{\mathcal{S}_{n+m}}$$

where χ and ψ are characters.

Before proving that this product agrees with the one in Λ, we must generalize some of the concepts we have previously introduced. Let G be any group and let $\mathcal{A}$ be any algebra over $\mathbf{C}$. Consider functions $\chi, \psi : G \to \mathcal{A}$ with the *bilinear form*

$$\langle \chi, \psi \rangle' = \frac{1}{|G|} \sum_{g \in G} \chi(g)\psi(g^{-1})$$

Note that since g and g^{-1} are in the same conjugacy class in $\mathcal{S}_n$, we have, for any class function χ,

$$\mathrm{ch}^n(\chi) = \frac{1}{n!} \sum_{\pi \in \mathcal{S}_n} \chi(\pi)p_\pi = \langle \chi, p \rangle'$$

where $p : \mathcal{S}_n \to \Lambda^n$ is the function $p(\pi) = p_\pi$.

Suppose $H \leq G$. If $\chi : G \to \mathcal{A}$, then define the *restriction of χ to H* to be the map $\chi\!\downarrow_H : H \to \mathcal{A}$ such that

$$\chi\!\downarrow_H (h) = \chi(h)$$

for all $h \in H$. On the other hand, if $\psi : H \to \mathcal{A}$, then let the *induction of ψ to G* be $\psi\!\uparrow^G : G \to \mathcal{A}$ defined by

$$\psi\!\uparrow^G (g) = \frac{1}{|H|} \sum_{x \in G} \psi(x^{-1}gx)$$

where $\psi = 0$ outside of H. The reader should verify that the following generalization of Frobenius reciprocity (Theorem 1.12.6) holds in this setting.

Theorem 4.7.3 *Consider $H \leq G$ with functions $\chi : G \to \mathcal{A}$ and $\psi : H \to \mathcal{A}$. If χ is a class function on G, then*

$$\langle \psi\!\uparrow^G, \chi \rangle' = \langle \psi, \chi\!\downarrow_H \rangle' \quad \blacksquare$$

This is the tool we need to get at our main theorem about the characteristic map.

Theorem 4.7.4 *The map $\mathrm{ch} : R \to \Lambda$ is an isomorphism of algebras.*

Proof. By Proposition 4.7.2, it suffices to check that products are preserved. If χ, ψ are characters in $\mathcal{S}_n$ and $\mathcal{S}_m$, respectively, then using part 2 of Therorem 1.11.2,

$$
\begin{aligned}
\mathrm{ch}(\chi \cdot \psi) &= \langle \chi \cdot \psi, p \rangle' \\
&= \langle (\chi \otimes \psi) \uparrow^{\mathcal{S}_{n+m}}, p \rangle' \\
&= \langle \chi \otimes \psi, p \downarrow_{\mathcal{S}_n \times \mathcal{S}_m} \rangle' \\
&= \frac{1}{n!m!} \sum_{\pi\sigma \in \mathcal{S}_n \times \mathcal{S}_m} (\chi \otimes \psi)(\pi\sigma) p_{\pi\sigma} \\
&= \frac{1}{n!m!} \sum_{\substack{\pi \in \mathcal{S}_n \\ \sigma \in \mathcal{S}_m}} \chi(\pi)\psi(\sigma) p_\pi p_\sigma \\
&= \left[\frac{1}{n!} \sum_{\pi \in \mathcal{S}_n} \chi(\pi) p_\pi \right] \left[\frac{1}{m!} \sum_{\sigma \in \mathcal{S}_m} \psi(\sigma) p_\sigma \right] \\
&= \mathrm{ch}(\chi)\,\mathrm{ch}(\psi) \ \blacksquare
\end{aligned}
$$

4.8 Knuth's Algorithm

Schensted [Sch 61] realized that his algorithm could be generalized to the case where the first output tableau is semistandard. Knuth [Knu 70] took this one step further, showing how to get a Robinson-Schensted map when both tableaux allow repetitions and thus making connection with the Cauchy identity [Lit 50, p. 103]. Many of the properties of the original algorithm are preserved. In addition, one obtains a new procedure that is dual to the first.

Just as we are to allow repetitions in our tableaux, we must also permit them in our permutations.

Definition 4.8.1 A *generalized permutation* is a two-line array of positive integers

$$
\pi = \begin{matrix} i_1 & i_2 & \cdots & i_n \\ j_1 & j_2 & \cdots & j_n \end{matrix}
$$

whose columns are in lexicographic order, with the top entry taking precedence. Let $\hat{\pi}$ and $\check{\pi}$ stand for the top and bottom rows of π, respectively. The set of all generalized permutations is denoted GP. $\blacksquare$

Note that the lexicographic condition can be restated: $\hat{\pi}$ is weakly increasing and in $\check{\pi}$, $j_k \leq j_{k+1}$ whenever $i_k = i_{k+1}$. An example of a generalized permutation is

$$
\pi = \begin{matrix} 1 & 1 & 1 & 2 & 2 & 3 \\ 2 & 3 & 3 & 1 & 2 & 1 \end{matrix} \tag{4.24}
$$

If T is any tableau, then let cont T be the content of T. If $\pi \in$ GP, then $\hat{\pi}$ and $\check{\pi}$ can be viewed as (one-rowed) tableaux, so their contents are defined. In the preceding example cont $\hat{\pi} = (3, 2, 1)$ and cont $\check{\pi} = (2, 2, 2)$.

The Robinson-Schensted-Knuth correspondence is as follows.

Theorem 4.8.2 ([Knu 70]) *There is a bijection between generalized permutations and pairs of semistandard tableaux of the same shape,*

$$\pi \stackrel{\text{R–S–K}}{\longleftrightarrow} (T, U)$$

such that cont $\check{\pi}$ = cont T *and* cont $\hat{\pi}$ = cont U.

Proof. $\pi \stackrel{\text{R–S–K}}{\longrightarrow} (T, U)$. We form, as before, a sequence of tableaux pairs

$$(T_0, U_0) = (\phi, \phi), \ (T_1, U_1), \ (T_2, U_2), \ \ldots, \ (T_n, U_n) = (T, U)$$

where the elements of $\check{\pi}$ are inserted into the T's and the elements of $\hat{\pi}$ are placed in the U's. The rules of insertion and placement are exactly the same as for tableaux without repetitions. Applying this algorithm to the preceding permutation, we obtain

$$
\begin{array}{cccccccc}
 & \phi, & 2, & 2\,3, & 2\,3\,3, & 1\,3\,3, & 1\,2\,3, & 1\,1\,3 \\
T_i: & & & & 2 & 2\,3 & 2\,2 & = T \\
 & & & & & & & 3 \\
\end{array}
$$

$$
\begin{array}{cccccccc}
 & \phi, & 1, & 1\,1, & 1\,1\,1, & 1\,1\,1, & 1\,1\,1, & 1\,1\,1 \\
U_i: & & & & 2 & 2\,2 & 2\,2 & = U \\
 & & & & & & & 3 \\
\end{array}
$$

It is easy to verify that the insertion rules ensure semistandardness of T. Also, U has weakly increasing rows because $\hat{\pi}$ is weakly increasing. To show that U's columns strictly increase, we must make sure that no two equal elements of $\hat{\pi}$ can end up in the same column. But if $i_k = i_{k+1} = i$ in the upper row, then we must have $j_k \leq j_{k+1}$. This implies that the insertion path for j_{k+1} will always lie strictly to the right of the path for j_k, which implies the desired result. Note that we have shown that all elements equal to i are placed in U from left to right as the algorithm proceeds.

$(T, U) \stackrel{\text{K–S–R}}{\longrightarrow} \pi$. Proceed as in the standard case. One problem is deciding which of the maximum elements of U corresponds to the last insertion. But from the observation just made, the rightmost of these maxima is the correct choice with which to start the deletion process.

We also need to verify that the elements removed from T corresponding to equal elements in U come out in weakly decreasing order. This is an easy exercise left to the reader. ∎

Knuth's original formulation of this algorithm had a slightly different point of departure. Just as one can consider a permutation as a matrix of zeros and ones, generalized permutations can be viewed as matrices with nonnegative integral entries. Specifically, with each $\pi \in GP$, we can associate a matrix of $M = M(\pi)$ with (i, j) entry

$$M_{i,j} = \text{the number of times } \binom{i}{j} \text{ occurs as a column of } \pi$$

Our example permutation has

$$M(\pi) = \begin{pmatrix} 0 & 1 & 2 \\ 1 & 1 & 0 \\ 1 & 0 & 0 \end{pmatrix}$$

We can clearly reverse this process, so we have a bijection

$$\pi \longleftrightarrow M$$

between GP and Mat = Mat($\mathbf{N}$), the set of all matrices with nonnegative integral entries and no final rows or columns of zeros (which contribute nothing to π). Note that in translating from π to M, the number of i's in $\hat{\pi}$ (respectively, the number of j's in $\check{\pi}$) becomes the sum of row i (respectively, column j) of M. Thus Theorem 4.8.2 can be restated.

Corollary 4.8.3 *There is a bijection between matrices $M \in$ Mat and pairs of semistandard tableaux of the same shape,*

$$M \stackrel{\mathrm{R-S-K}}{\longleftrightarrow} (T, U)$$

such that cont T *and* cont U *give the vectors of column and row sums, respectively, of* M. ∎

To translate these results into generating functions, we use two sets of variables, $\mathbf{x} = \{x_1, x_2, \ldots\}$ and $\mathbf{y} = \{y_1, y_2, \ldots\}$. Let the weight of a generalized permutation π be

$$\mathrm{wt}\,\pi = \mathbf{x}^{\hat{\pi}}\mathbf{y}^{\check{\pi}} = \mathbf{x}^{\mathrm{cont}\,\hat{\pi}}\mathbf{y}^{\mathrm{cont}\,\check{\pi}}$$

To illustrate, the permutation in (4.24) has

$$\mathrm{wt}\,\pi = x_1^3 x_2^2 x_3 y_1^2 y_2^2 y_3^2$$

If the column $\begin{pmatrix} i \\ j \end{pmatrix}$ occurs k times in π, then it gives a contribution of $x_i^k y_j^k$ to the weight. Thus the generating function for generalized permutations is

$$\sum_{\pi \in \mathrm{GP}} \mathrm{wt}\,\pi = \prod_{i,j \geq 1} \sum_{k \geq 0} x_i^k y_j^k$$

$$= \prod_{i,j \geq 1} \frac{1}{1 - x_i y_j}$$

As for tableaux pairs (T, U), let

$$\mathrm{wt}(T, U) = \mathbf{x}^U \mathbf{y}^T = \mathbf{x}^{\mathrm{cont}\,U}\mathbf{y}^{\mathrm{cont}\,T}$$

as in equation (4.8). Restricting to pairs of the same shape,

$$\sum_{\mathrm{sh}\,T = \mathrm{sh}\,U} \mathrm{wt}(T, U) = \sum_{\lambda} \left(\sum_{\mathrm{sh}\,U = \lambda} \mathbf{x}^U \right) \left(\sum_{\mathrm{sh}\,T = \lambda} \mathbf{y}^T \right)$$

$$= \sum_{\lambda} s_\lambda(\mathbf{x}) s_\lambda(\mathbf{y})$$

Since the Robinson-Schensted-Knuth map is weight preserving, we have proved Cauchy's formula.

Theorem 4.8.4 ([Lit 50]) *We have*

$$\sum_\lambda s_\lambda(\mathbf{x}) s_\lambda(\mathbf{y}) = \prod_{i,j \geq 1} \frac{1}{1 - x_i y_j} \quad \blacksquare$$

Note that, just as the Robinson-Schensted correspondence is gotten by restricting Knuth's generalization to the case where all entries are distinct, we can obtain

$$n! = \sum_{\lambda \vdash n} (f^\lambda)^2$$

by taking the coefficient of $x_1 \cdots x_n y_1 \cdots y_n$ on both sides of Theorem 4.8.4.

Because the semistandard condition does not treat rows and columns uniformly, there is a second algorithm related to the one just given. It is called the *dual map*. (This is a different notion of "dual" from the one introduced in Chapter 3—e.g., Definition 3.8.8.) Let GP′ denote all those permutations in GP where no column is repeated. These correspond to the $0 - 1$ matrices in Mat.

Theorem 4.8.5 ([Knu 70]) *There is a bijection between $\pi \in$ GP′ and pairs (T, U) of tableaux of the same shape with T, U^t semistandard*

$$\pi \stackrel{R-S-K'}{\longleftrightarrow} (T, U)$$

such that cont $\check{\pi}$ = cont T *and* cont $\hat{\pi}$ = cont U.

Proof. $\pi \stackrel{\text{R-S-K}'}{\longrightarrow} (T, U)$. We merely replace row insertion in the R-S-K correspondence with a modification of column insertion. This is done by insisting that at each stage the element entering a column displaces the smallest entry greater than *or equal* to it. For example,

$$c_2 \begin{pmatrix} 1 & 1 & 3 \\ 2 & 3 \\ 3 \end{pmatrix} = \begin{matrix} 1 & 1 & 3 & 3 \\ 2 & 2 \\ 3 \end{matrix}$$

Note that this is exactly what is needed to ensure that T will be column-strict. The fact that U will be row-strict follows because a subsequence of $\check{\pi}$ corresponding to equal elements in $\hat{\pi}$ must be strictly increasing (since $\pi \in$ GP′).

$(T, U) \stackrel{\text{K-S-R}'}{\longrightarrow} \pi$. The details of the step-by-step reversal and verification that $\pi \in$ GP′ are routine. $\blacksquare$

Taking generating functions with the same weights as before yields the dual Cauchy identity.

Theorem 4.8.6 ([Lit 50]) *We have*

$$\sum_\lambda s_\lambda(\mathbf{x})s_{\lambda'}(\mathbf{y}) = \prod_{i,j \geq 1} (1 + x_i y_j)$$

where λ' is the conjugate of λ. ∎

Most of the results of Chapter 3 about Robinson-Schensted have generalizations for the Knuth map. We survey a few of them next.

Taking the inverse of a permutation corresponds to transposing the associated permutation matrix. So the following strengthening of Schützenberger's Theorem 3.8.6 should come as no surprise.

Theorem 4.8.7 *If $M \in$ Mat and $M \overset{\text{R-S-K}}{\longleftrightarrow} (T, U)$, then*

$$M^t \overset{\text{R-S-K}}{\longleftrightarrow} (U, T) \quad \blacksquare$$

We can also deal with the reversal of a generalized permutation. Row and modified column insertion commute as in Proposition 3.4.2. So we obtain the following analog of Theorem 3.4.3.

Theorem 4.8.8 *If $\pi \in$ GP, then $T(\check{\pi}^r) = T(\check{\pi})^t$.* ∎

The Knuth relations become

$$\text{replace } xzy \text{ by } zxy \text{ if } x \leq y < z$$

and

$$\text{replace } yxz \text{ by } yzx \text{ if } x < y \leq z$$

Theorem 3.6.3 remains true.

Theorem 4.8.9 ([Knu 70]) *Two generalized permutations are Knuth equivalent if and only if they have the same T-tableau.* ∎

Putting together the last two results, we can prove a stronger version of Greene's theorem.

Theorem 4.8.10 ([Gre 74]) *Given $\pi \in$ GP, let $\text{sh}\, T(\pi) = (\lambda_1, \lambda_2, \ldots, \lambda_l)$ with conjugate $(\lambda'_1, \lambda'_2, \ldots, \lambda'_m)$. Then for any k, $\lambda_1 + \lambda_2 + \cdots + \lambda_k$ and $\lambda'_1 + \lambda'_2 + \cdots + \lambda'_k$ give the lengths of the longest weakly k-increasing and strictly k-decreasing subsequences of π, respectively.* ∎

For the jeu de taquin, we need to break ties when the two elements of T adjacent to the cell to be filled are equal. The correct choice is forced on us by semistandardness. In this case, both the forward and reverse slides always move the element that changes rows rather than the one that would change columns. The fundamental results of Schützenberger continue to hold (see Theorems 3.9.7 and 3.9.8).

Theorem 4.8.11 ([Scü 76]) *Let T and U be skew semistandard tableaux. Then T and U have Knuth equivalent row words if and only if they are connected by a sequence of slides. Furthermore, any such sequence bringing them to normal shape results in the first output tableaux of the Robinson-Schensted-Knuth correspondence.* ∎

Finally, we can define dual equivalence, $\overset{*}{\cong}$, in exactly the same way as before (Definition 3.10.2). The result concerning this relation, analogous to Proposition 3.10.1 and Theorem 3.10.8, needed for the Littlewood-Richardson rule is the following.

Theorem 4.8.12 *If T and U are semistandard of the same normal shape, then $T \overset{*}{\cong} U$.* ∎

4.9 The Littlewood-Richardson Rule

The Littlewood-Richardson rule gives a combinatorial interpretation to the coefficients of the product $s_\mu s_\nu$ when expanded in terms of the Schur basis. This can be viewed as a generalization of Young's rule, as follows.

We know (Theorem 2.11.2) that

$$M^\mu \cong \bigoplus_\lambda K_{\lambda\mu} S^\lambda \tag{4.25}$$

where $K_{\lambda\mu}$ is the number of semistandard tableaux of shape λ and content μ. We can look at this formula from two other perspectives: in terms of characters or symmetric functions.

If $\mu \vdash n$, then M^μ is a module for the induced character $1_{S_\mu} \uparrow^{S_n}$. But from the definitions of the trivial character and the tensor product, we have

$$1_{S_\mu} = 1_{S_{\mu_1}} \otimes 1_{S_{\mu_2}} \otimes \cdots \otimes 1_{S_{\mu_m}}$$

where $\mu = (\mu_1, \mu_2, \ldots, \mu_m)$. Using the product in the class function algebra R (and the transitivity of induction, Exercise 18 of Chapter 1), we can rewrite (4.25) as

$$1_{S_{\mu_1}} \cdot 1_{S_{\mu_2}} \cdots 1_{S_{\mu_m}} = \sum_\lambda K_{\lambda\mu} \chi^\lambda$$

To bring in symmetric functions, apply the characteristic map to the previous equation. (Remember that the trivial representation corresponds to an irreducible whose diagram has only one row.)

$$s_{(\mu_1)} s_{(\mu_2)} \cdots s_{(\mu_m)} = \sum_\lambda K_{\lambda\mu} s_\lambda$$

For example,

$$M^{(3,2)} = S^{(3,2)} + S^{(4,1)} + S^{(5)}$$

with the relevant tableaux being

$$\begin{array}{lll} 1\ 1\ 1, & 1\ 1\ 1\ 2, & 1\ 1\ 1\ 2\ 2 \\ 2\ 2 & 2 & \end{array}$$

This can be rewritten as

$$1_{S_3} \cdot 1_{S_2} = \chi^{(3,2)} + \chi^{(4,1)} + \chi^{(5)}$$

or

$$s_{(3)}s_{(2)} = s_{(3,2)} + s_{(4,1)} + s_{(5)}$$

What happens if we try to compute the expansion

$$s_\mu s_\nu = \sum_\lambda c_{\mu\nu}^\lambda s_\lambda \tag{4.26}$$

where μ and ν are arbitrary partitions? Equivalently, we are asking for the multiplicities of the irreducibles in

$$\chi^\mu \cdot \chi^\nu = \sum_\lambda c_{\mu\nu}^\lambda \chi^\lambda$$

or

$$(S^\mu \otimes S^\nu)\uparrow^{S_n} = \bigoplus_\lambda c_{\mu\nu}^\lambda S^\lambda$$

where $|\mu| + |\nu| = n$. The $c_{\mu\nu}^\lambda$ are called the *Littlewood-Richardson coefficients*. The importance of the Littlewood-Richardson rule below is that it gives a way to interpret these coefficients combinatorially, just as Young's rule does for one-rowed partitions.

We need to explore one other place where these coefficients arise: in the expansion of *skew Schur functions*. Obviously, the definition of $s_\lambda(\mathbf{x})$ given in Section 4.4 makes sense if λ is replaced by a skew diagram. Furthermore, the resulting function $s_{\lambda/\mu}(\mathbf{x})$ is still symmetric by the same reasoning as in Proposition 4.4.2. We can derive an implicit formula for these new Schur functions in terms of the old ones by introducing another set of indeterminates $\mathbf{y} = (y_1, y_2, \ldots)$.

Proposition 4.9.1 ([Mac 79]) *Define* $s_\lambda(\mathbf{x}, \mathbf{y}) = s_\lambda(x_1, x_2, \ldots, y_1, y_2, \ldots)$. *Then*

$$s_\lambda(\mathbf{x}, \mathbf{y}) = \sum_{\mu \subseteq \lambda} s_\mu(\mathbf{x}) s_{\lambda/\mu}(\mathbf{y}) \tag{4.27}$$

Proof. The function $s_\lambda(\mathbf{x}, \mathbf{y})$ enumerates semistandard fillings of the diagram λ with letters from the totally ordered alphabet

$$\{1 < 2 < 3 < \cdots < 1' < 2' < 3' < \cdots\}$$

In any such tableaux, the unprimed numbers (which are weighted by the x's) form a subtableaux of shape μ in the upper left corner of λ, whereas the primed numbers (weighted by the y's) fill the remaining squares of λ/μ. The right-hand side of (4.27) is the generating function for this description of the relevant tableaux. ∎

Since $s_{\lambda/\mu}(\mathbf{x})$ is symmetric, we can express it as a linear combination of ordinary Schur functions. Some familiar coefficients will then appear.

Theorem 4.9.2 *If the $c_{\mu\nu}^{\lambda}$ are Littlewood-Richardson coefficients, where $|\mu|+|\nu|=|\lambda|$, then*

$$s_{\lambda/\mu} = \sum_{\nu} c_{\mu\nu}^{\lambda} s_{\nu}.$$

Proof. Bring in yet a third set of variables $\mathbf{z} = \{z_1, z_2, \ldots\}$. By using the previous proposition and Cauchy's formula (Theorem 4.8.4),

$$
\begin{aligned}
\sum_{\lambda,\mu} s_\mu(\mathbf{x})s_{\lambda/\mu}(\mathbf{y})s_\lambda(\mathbf{z}) &= \sum_\lambda s_\lambda(\mathbf{x},\mathbf{y})s_\lambda(\mathbf{z}) \\
&= \prod_{i,j} \frac{1}{1-x_i z_j}\frac{1}{1-y_i z_j} \\
&= \left[\sum_\mu s_\mu(\mathbf{x})s_\mu(\mathbf{z})\right]\left[\sum_\nu s_\nu(\mathbf{y})s_\nu(\mathbf{z})\right] \\
&= \sum_{\mu,\nu} s_\mu(\mathbf{x})s_\nu(\mathbf{y})s_\mu(\mathbf{z})s_\nu(\mathbf{z})
\end{aligned}
$$

Taking the coefficient of $s_\mu(\mathbf{x})s_\nu(\mathbf{y})s_\lambda(\mathbf{z})$ on both sides and comparing with equation (4.26) completes the proof. ∎

One last definition is needed to explain what the $c_{\mu\nu}^{\lambda}$ count.

Definition 4.9.3 A *ballot sequence* or *lattice permutation* is a sequence of positive integers $\pi = i_1 i_2 \ldots i_n$ such that, for any prefix $\pi_k = i_1 i_2 \ldots i_k$ and any positive integer l, the number of l's in π_k is at least as large as the number of $l+1$s in that prefix. A *reverse* ballot sequence or lattice permutation is π such that π^r is a ballot sequence. ∎

As an example,

$$\pi = 1\ 1\ 2\ 3\ 2\ 1\ 3$$

is a lattice permutation, wereas

$$\pi = 1\ 2\ 3\ 2\ 1\ 1\ 3$$

is not because the prefix 1 2 3 2 has more twos than ones.

The name *ballot sequence* comes from the following scenario. Suppose the ballots from an election are being counted sequentially. Then a ballot sequence corresponds to a counting where candidate one always (weakly) leads candidate two, candidate two always (weakly) leads candidate three, and so on.

Furthermore, lattice permutations are just another way of encoding standard tableaux. Given P standard with n elements, form the sequence $\pi = i_1 i_2 \ldots i_n$, where $i_k = i$ if k appears in row i of P. The fact that the entries less than or equal to k form a partition with weakly decreasing parts translates to the ballot condition on π_k. It is easy to construct the inverse correspondence and see that our example lattice permutation codes the tableau

$$P = \begin{array}{ccc} 1 & 2 & 6 \\ 3 & 5 & \\ 4 & 7 & \end{array}$$

We are finally ready to state and prove the Littlewood-Richardson rule. Although it was first stated by these two authors [L-R 34], complete proofs were not published until comparatively recently by Thomas [Tho 74, Tho 78] and Schützenberger [Scü 76]; our demonstration is based on the latter.

Theorem 4.9.4 (Littlewood-Richardson Rule [L-R 34]) *The value of the coefficient $c_{\mu\nu}^{\lambda}$ is equal to the number of semistandard tableaux T such that*

1. *T has shape λ/μ and content ν,*

2. *the row word of T, π_T, is a reverse lattice permutation.*

Proof. Let $d_{\mu\nu}^{\lambda}$ be the number of tableaux T satisfying the restrictions of the theorem. Then we claim that it suffices to find a weight-preserving map

$$T \xrightarrow{j} U \tag{4.28}$$

from semistandard tableaux T of shape λ/μ to semistandard tableaux U of normal shape such that

1. the number of tableaux T mapping to a given U of shape ν depends only on λ, μ, and ν and not on U itself;

2. the number of tableaux T mapping to a particular choice of U is $d_{\mu\nu}^{\lambda}$.

If such a bijection exists, then

$$s_{\lambda/\mu}(\mathbf{x}) = \sum_{\mathrm{sh}\, T = \lambda/\mu} \mathbf{x}^T = \sum_{\nu} d_{\mu\nu}^{\lambda} \sum_{\mathrm{sh}\, U = \nu} \mathbf{x}^U = \sum_{\nu} d_{\mu\nu}^{\lambda} s_{\nu}(\mathbf{x})$$

Comparison of this expression with Theorem 4.9.2 and the fact that the Schur functions are a basis completes the proof.

It turns out that the map j needed for equation (4.28) is just the jeu de taquin! To show that j satisfies property 1, consider U and U' of the same shape ν. We will define a bijection between the set of tableaux T that map to U and the set of T' that map to U' as follows. Let P be a fixed standard tableau of shape μ. Then, in the notation of Section 3.10, we always have

$U = j^P(T)$. Let the vacating tableaux for this sequence of slides be denoted by Q_T, a standard tableaux of shape λ/ν. Consider the composition

$$T \xrightarrow{j^P} U \longrightarrow U' \xrightarrow{j_{Q_T}} T' \tag{4.29}$$

—i.e., slide T to normal shape to obtain U and Q_T, replace U by U', and then slide U' back out again according to the evacuating tableaux.

We first verify that this function is well defined in that T' must be sent to U' by jeu de taquin and must have shape λ/μ. The first statement is clear, since all the slides in j_{Q_T} are reversible. For the second assertion, note that by the definition of the vacating tableau,

$$T = j_{Q_T} j^P(T) = j_{Q_T}(U)$$

Also,

$$T' = j_{Q_T}(U')$$

Since U and U' are of the same normal shape, they are dual equivalent (Theorem 4.8.12), which implies that

$$T \stackrel{*}{\cong} T' \tag{4.30}$$

In particular, T and T' must have the same shape, as desired.

To show that (4.29) is a bijection, we claim that

$$T' \xrightarrow{j^P} U' \longrightarrow U \xrightarrow{j_{Q_{T'}}} T$$

is its inverse, where $Q_{T'} = v^P(T')$. But equation (4.30) implies that the corresponding vacating tableaux Q_T and $Q_{T'}$ are equal. So

$$j^P(T') = j^P j_{Q_T}(U') = j^P j_{Q_{T'}}(U') = U'$$

and

$$j_{Q_{T'}}(U) = j_{Q_{T'}} j^P(T) = j_{Q_T} j^P(T) = T$$

showing that our map is a well-defined inverse.

Now we must choose a particular tableaux U_0 for property 2. Let U_0 be the superstandard tableau of shape and content ν, where the elements of the ith row are all i's. Notice that the row word π_{U_0} is a reverse lattice permutation. In considering all tableaux mapped to U_0 by j, it is useful to have the following result.

Lemma 4.9.5 *If we can go from T to T' by a sequence of slides, then π_T is a reverse lattice permutation if and only if $\pi_{T'}$ is.*

Proof. It suffices to consider the case where T and T' differ by a single move. If the move is horizontal, the row word doesn't change, so consider a vertical

move. Suppose

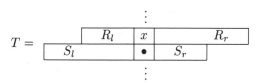

and

$$T' =$$

where R_l and S_l (respectively, R_r and S_r) are the left (respectively, right) portions of the two rows between which x is moving. We show that if $\sigma = \pi_T^r$ is a ballot sequence, then so is $\sigma' = \pi_{T'}^r$. (The proof of the reverse implication is similar.)

Clearly we only need to check that the number of $x + 1$s does not exceed the number of x's in any prefix of σ' ending with an element of R_l or S_r. To do this, we show that each $x + 1$ in such a prefix can be injectively matched with an x that comes before it. If the $x + 1$ occurs in R_r or a higher row, then it can be matched with an x because σ is a ballot sequence. Notice that all these x's must be in higher rows because the x's in R_r come after the $x + 1$s in R_r when listed in the reverse row word. By semistandardness, there are no $x + 1$s in R_l to be matched. For the same reason, every $x + 1$ in S_r must have an x in R_r just above it. Since these x's have not been previously used in our matching, we are done. ∎

Now note that U_0 is the unique semistandard tableau of normal shape ν whose row word is lattice permutation. Thus by the lemma, the set of tableaux T with shape λ/μ such that $j(T) = U_0$ are precisely those described by the coefficients $d_{\mu\nu}^\lambda$. This finishes the proof of the Littlewood-Richardson rule. ∎

As an illustration, we can calculate the product $s_{(2,1)}s_{(2,2)}$. Listing all tableaux subject to the ballot sequence condition with content $(2,2)$ and skew shape $\lambda/(2,1)$ for some λ yields

$$
\begin{array}{l}
\begin{array}{ccc} \bullet & \bullet & 1 \ 1 \\ \bullet & 2 & 2 \end{array}, \quad
\begin{array}{ccc} \bullet & \bullet & 1 \ 1 \\ \bullet & 2 \\ 2 \end{array}, \quad
\begin{array}{ccc} \bullet & \bullet & 1 \\ \bullet & 1 & 2 \\ 2 \end{array}, \quad
\begin{array}{ccc} \bullet & \bullet & 1 \\ \bullet & 1 \\ 2 & 2 \end{array}, \quad
\begin{array}{ccc} \bullet & \bullet & 1 \\ \bullet & 2 \\ 1 \\ 2 \end{array}, \quad
\begin{array}{ccc} \bullet & \bullet \\ \bullet & 1 \\ 1 & 2 \\ 2 \end{array}
\end{array}
$$

Thus

$$s_{(2,1)}s_{(2,2)} = s_{(4,3)} + s_{(4,2,1)} + s_{(3^2,1)} + s_{(3,2^2)} + s_{(3,2,1^2)} + s_{(2^3,1)}$$

For another example, let's find the coefficient of $s_{(5,3,2,1)}$ in $s_{(3,2,1)}s_{(3,2)}$. First we fix the outer shape $\lambda = (5,3,2,1)$ and then find the tableaux as before

$$
\begin{array}{ccccc}
\bullet & \bullet & \bullet & 1 & 1, \\
\bullet & \bullet & 1 & & \\
\bullet & 2 & & & \\
2 & & & & \\
\end{array}
\qquad
\begin{array}{ccccc}
\bullet & \bullet & \bullet & 1 & 1, \\
\bullet & \bullet & 2 & & \\
\bullet & 1 & & & \\
2 & & & & \\
\end{array}
\qquad
\begin{array}{ccccc}
\bullet & \bullet & \bullet & 1 & 1 \\
\bullet & \bullet & 2 & & \\
\bullet & 2 & & & \\
1 & & & & \\
\end{array}
$$

It follows that

$$c_{(3,2,1)(3,2)}^{(5,3,2,1)} = 3$$

4.10   The Murnaghan-Nakayama Rule

The Murnaghan-Nakayama rule [Mur 37, Nak 40] is a combinatorial way of computing the value of the irreducible character χ^λ on the conjugacy class α. The crucial objects that come into play are the skew hooks.

Definition 4.10.1 A *skew hook*, or *rim hook*, ξ, is a skew diagram obtained by taking all cells on a finite lattice path with steps one unit north or east. Equivalently, ξ is a skew hook if it is edgewise connected and contains no 2×2 subset of cells:

$$\qquad\qquad\qquad\qquad\qquad\qquad\qquad\qquad (4.31)$$

The *leg length* of ξ is

$$ll(\xi) = (\text{the number of rows of } \xi) \; -1 \; \blacksquare$$

For example,

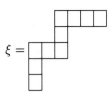

$$\xi =$$

has $ll(\xi) = 4$. The name *rim hook* comes from the fact that such a diagram can be obtained by projecting a regular hook along diagonals onto the boundary of a shape. Our example hook is the projection of $H_{1,1}$ onto the rim of $\lambda = (6,3,3,1,1)$.

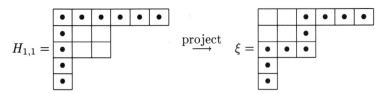

$$H_{1,1} = \qquad\qquad \xrightarrow{\text{project}} \qquad \xi =$$

Notice that we are using single Greek letters such as ξ for rim hooks even though they are skew. If $\xi = \lambda/\mu$, then we write $\lambda \backslash \xi$ for μ. In the preceding example, $\mu = (2,2)$. Also, if $\alpha = (\alpha_1, \alpha_2, \ldots, \alpha_k)$ is a composition, then let $\alpha \backslash \alpha_1 = (\alpha_2, \ldots, \alpha_k)$. With this notation, we can state the main result of this section.

Theorem 4.10.2 (Murnaghan-Nakayama Rule [Mur 37, Nak 40]) *If λ is a partition of n and $\alpha = (\alpha_1, \ldots, \alpha_k)$ is a composition of n, then we have*

$$\chi_\alpha^\lambda = \sum_\xi (-1)^{ll(\xi)} \chi_{\alpha \backslash \alpha_1}^{\lambda \backslash \xi}$$

where the sum runs over all rim hooks ξ of λ having α_1 cells.

Before proving this theorem, a few remarks are in order. To calculate χ_α^λ the rule must be used iteratively. First remove a rim hook from λ with α_1 cells in all possible ways such that what's left is a normal shape. Then strip away hooks with α_2 squares from the resulting diagrams, and so on. At some stage either it will be impossible to remove a rim hook of the right size (so the contribution of the corresponding character is zero) or all cells will be deleted (giving a contribution of $\pm \chi_{(0)}^{(0)} = \pm 1$).

To illustrate the process, we compute $\chi_{(5,4,2)}^{(4,4,3)}$. The stripping of hooks can be viewed as a tree. Cells to be removed are marked with a dot and the appropriate sign appears to the right of the diagram.

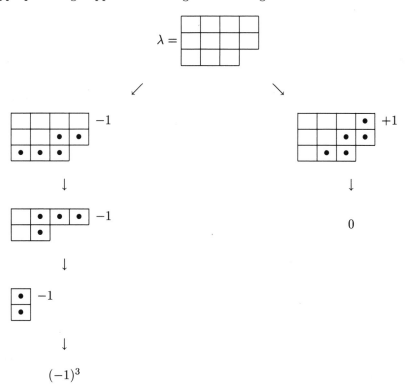

The corresponding calculations are

$$
\begin{aligned}
\chi^{(4,4,3)}_{(5,4,2)} &= -\chi^{(4,2)}_{(4,2)} + \chi^{(3,2,1)}_{(4,2)} \\
&= -(-\chi^{(1,1)}_{(2)}) + 0 \\
&= -(-(-\chi^{(0)}_{(0)})) \\
&= (-1)^3
\end{aligned}
$$

Also note that the branching rule from Section 2.8 is a special case of Murnaghan-Nakayama. Take $\alpha = (1, \alpha_2, \ldots, \alpha_k)$ and let $\pi \in \mathcal{S}_n$ have type α. Since π has a fixed-point,

$$
\chi^\lambda_\alpha = \chi^\lambda(\pi) = \chi^\lambda\!\downarrow_{\mathcal{S}_{n-1}} (\pi)
$$

which corresponds to the left-hand side of the first equation in Theorem 2.8.3. As for the right side, $|\xi| = 1$ forces $\lambda \backslash \xi$ to be of the form λ^- with all signs $(-1)^0 = +1$.

We now embark on a proof of the Murnaghan-Nakayama rule. It is based on the one in James [Jam 78, pp. 79–83].

Proof (of Theorem 4.10.2). Let $m = \alpha_1$. Consider $\pi\sigma \in \mathcal{S}_{n-m} \times \mathcal{S}_m \subseteq \mathcal{S}_n$, where π has type $(\alpha_2, \ldots, \alpha_k)$ and σ is an m-cycle. By part 2 of Theorem 1.11.3, the characters $\chi^\mu \otimes \chi^\nu$, where $\mu \vdash n - m$, $\nu \vdash m$, form a basis for the class functions on $\mathcal{S}_{n-m} \times \mathcal{S}_m$. So

$$
\chi^\lambda_\alpha = \chi^\lambda(\pi\sigma) = \chi^\lambda\!\downarrow_{\mathcal{S}_{n-m} \times \mathcal{S}_m} (\pi\sigma) = \sum_{\substack{\mu \vdash n-m \\ \nu \vdash m}} m^\lambda_{\mu\nu} \chi^\mu(\pi)\chi^\nu(\sigma) \qquad (4.32)
$$

To find the multiplicities, we use Frobenius reciprocity (Theorem 1.12.6) and the characteristic map.

$$
\begin{aligned}
m^\lambda_{\mu\nu} &= \langle \chi^\lambda\!\downarrow_{\mathcal{S}_{n-m} \times \mathcal{S}_m}, \chi^\mu \otimes \chi^\nu \rangle \\
&= \langle \chi^\lambda, (\chi^\mu \otimes \chi^\nu)\!\uparrow^{\mathcal{S}_n} \rangle \\
&= \langle \chi^\lambda, \chi^\mu \cdot \chi^\nu \rangle \\
&= \langle s_\lambda, s_\mu s_\nu \rangle \\
&= c^\lambda_{\mu\nu}
\end{aligned}
$$

where $c^\lambda_{\mu\nu}$ is a Littlewood-Richardson coefficient. Thus we can write equation (4.32) as

$$
\chi^\lambda(\pi\sigma) = \sum_{\mu \vdash n-m} \chi^\mu(\pi) \sum_{\nu \vdash m} c^\lambda_{\mu\nu} \chi^\nu(\sigma) \qquad (4.33)
$$

Now we must evaluate $\chi^\nu(\sigma)$, where σ is an m-cycle.

Lemma 4.10.3 *If $\nu \vdash m$, then*

$$
\chi^\nu_{(m)} = \begin{cases} (-1)^{m-r} & \text{if } \nu = (r, 1^{m-r}) \\ 0 & \text{otherwise} \end{cases}
$$

Note that this is a special case of the theorem we are proving. If $\alpha = (m)$, then $\chi_\alpha^\nu \neq 0$ only when we can remove all the cells of ν in a single sweep—that is, when ν itself is a hook diagram $(r, 1^{m-r})$. In this case, the Murnaghan-Nakayama sum has a single term with leg length $m - r$.

Proof (of Lemma 4.10.3). By equation (4.23), $\chi_{(m)}^\nu$ is $z_{(m)} = m$ times the coefficient of p_m in

$$s_\nu = \sum_\mu \frac{1}{z_\mu} \chi_\mu^\nu p_\mu$$

Using the complete homogeneous Jacobi-Trudi determinant (Theorem 4.5.1),

$$s_\nu = |h_{\nu_i - i + j}|_{l \times l} = \sum_\kappa \pm h_\kappa$$

where the sum is over all compositions $\kappa = (\kappa_1, \ldots, \kappa_l)$ that occur as a term in the determinant. But each h_{κ_i} in h_κ can be written as a linear combination of power sums. So, since the p's are a multiplicative basis, the resulting linear combination for h_κ will not contain p_m unless κ contains exactly one nonzero part which must, of course, be m. Hence $\chi_{(m)}^\nu \neq 0$ only when h_m appears in the preceding determinant.

The largest index to appear in this determinant is at the end of the first row, and $\nu_1 - 1 + l = h_{1,1}$, the hooklength of cell $(1,1)$. Furthermore, we always have $m = |\nu| \geq h_{1,1}$. Thus $\chi_{(m)}^\nu$ is nonzero only when $h_{1,1} = m$—i.e., when ν is a hook $(r, 1^{m-r})$. In this case, we have

$$s_\nu = \begin{vmatrix} h_r & \cdots & & & & h_m \\ h_0 & h_1 & \cdots & & & \\ 0 & h_0 & h_1 & \cdots & & \\ 0 & 0 & h_0 & h_1 & \cdots & \\ \vdots & \vdots & \vdots & \vdots & \vdots & \vdots \end{vmatrix}$$

$$= (-1)^{m-r} h_m + \text{other terms not involving } p_m$$

But $h_m = s_{(m)}$ corresponds to the trivial character, so comparing coefficients of p_m/m in this last set of equalities yields $\chi_{(m)}^\nu = (-1)^{m-r}$, as desired. ∎

In view of this lemma and equation (4.33), we need to find $c_{\mu\nu}^\lambda$ when ν is a hook.

Lemma 4.10.4 *Let $\lambda \vdash n$, $\mu \vdash n - m$ and $\nu = (r, 1^{m-r})$. Then $c_{\mu\nu}^\lambda = 0$ unless each edgewise connected component of λ/μ is a rim hook. In that case, if there are k component hooks spanning a total of c columns, then*

$$c_{\mu\nu}^\lambda = \binom{k-1}{c-r}$$

Proof. By the Littlewood-Richardson rule (Theorem 4.9.4), $c_{\mu\nu}^\lambda$ is the number of semistandard tableaux T of shape λ/μ containing r ones and a single copy each of $2, 3, \ldots, m - r + 1$ such that π_T is a reverse lattice permutation. Thus the numbers greater than one in π_T^r must occur in increasing order. This condition, together with semistandardness, puts the following constraints on T.

T1. Any cell of T having a cell to its right must contain a one.

T2. Any cell of T having a cell above must contain an element bigger than one.

If T contains a 2×2 block of squares, as in (4.31), then there is no way to fill the lower left cell and satisfy both T1 and T2. Thus $c_{\mu\nu}^\lambda = 0$ if the components of the shape of T are not rim hooks.

Now suppose $\lambda/\mu = \biguplus_{i=1}^k \xi^{(i)}$, where each $\xi^{(i)}$ is a component skew hook. Conditions T1, T2 and the fact that 2 through $m - r + 1$ increase in π_T^r show that every rim hook must have the form

$$
\xi^{(i)} \;=\; \begin{array}{|c|c|c|c|}
\hline
 & 1 & 1 & 1 & b \\
\cline{2-5}
 & d & \multicolumn{3}{c}{} \\
\cline{2-2}
\multicolumn{1}{c|}{1} & 1 & d+1 & \multicolumn{2}{c}{} \\
\cline{1-3}
\multicolumn{1}{|c|}{d+2} & \multicolumn{4}{c}{} \\
\cline{1-1}
\multicolumn{1}{|c|}{d+3} & \multicolumn{4}{c}{} \\
\cline{1-1}
\end{array}
$$

where $d > 1$ is the smallest number that has not yet appeared in π_T^r and b is either 1 or $d - 1$. Thus all the entries in $\xi^{(i)}$ are determined once we choose the value of b. Furthermore, in $\xi^{(1)}$ we must have $b = 1$. By T2 we have $c - (k - 1)$ ones already fixed in T. Hence there are $r - c + k - 1$ ones left to distribute among the $k - 1$ cells marked with a b. The number of ways this can be done is

$$
c_{\mu\nu}^\lambda = \binom{k-1}{r-c+k-1} = \binom{k-1}{c-r} \quad \blacksquare
$$

Putting the values from the two lemmas in equation (4.33) we obtain

$$
\chi^\lambda(\pi\sigma) = \sum_\mu \chi^\mu(\pi) \sum_{r=1}^m \binom{k-1}{c-r}(-1)^{m-r} \tag{4.34}
$$

Now $k \le c \le m$, since these three quantities represent the number of skew hooks $\xi^{(i)}$, the number of columns in the $\xi^{(i)}$, and the number of cells in the $\xi^{(i)}$, respectively. Thus

$$
\sum_{r=1}^m \binom{k-1}{c-r}(-1)^{m-r} = (-1)^{m-c}\left[\binom{k-1}{0} - \binom{k-1}{1} + \cdots \pm \binom{k-1}{k-1}\right]
$$

$$
= \begin{cases} (-1)^{m-c} & \text{if } k-1 = 0 \\ 0 & \text{otherwise} \end{cases}
$$

But if $k = 1$, then λ/μ is a single skew hook ξ with m squares and c columns. Hence $m - c = ll(\xi)$, so equation (4.34) becomes

$$
\chi^\lambda(\pi\sigma) = \sum_{|\xi|=m} (-1)^{ll(\xi)} \chi^{\lambda\backslash\xi}(\pi)
$$

finishing the proof of the Murnaghan-Nakayama rule. ∎

It is possible to combine the calculations for χ_α^λ into single tableaux.

Definition 4.10.5 A *rim hook tableau* is a generalized tableau T with positive integral entries such that

1. rows and columns of T weakly increase, and

2. all occurrences of i in T lie in a single rim hook. ∎

Obviously the stripping of rim hooks from a diagram determines a rim hook tableau by labeling the last hook removed with ones, the next to last with twos, and so on. For example, the tableau corresponding to the left branch of the preceding tree is

$$T = \begin{array}{cccc} 1 & 2 & 2 & 2 \\ 1 & 2 & 3 & 3 \\ 3 & 3 & 3 \end{array}$$

Now define the *sign* of a rim hook tableau with rim hooks $\xi^{(i)}$ to be

$$(-1)^T = \prod_{\xi^{(i)} \in T} (-1)^{ll(\xi^{(i)})}$$

The tableaux just given has $\text{sgn}(T) = (-1)^3$. It is easy to see that the Murnaghan-Nakayama theorem can be restated.

Corollary 4.10.6 *Let λ be a partition of n and let $\alpha = (\alpha_1, \dots, \alpha_k)$ be any composition of n. Then*

$$\chi_\alpha^\lambda = \sum_T (-1)^T$$

where the sum is over all rim hook tableaux of shape λ and content α. ∎

We should mention that Stanton and White [S-W 85] have extended Schensted's construction to rim hook tableaux. This permitted White [Whi 83] to give a bijective proof of the orthonormality of the irreducible characters of S_n (Theorem 1.9.3) analogous to the combinatorial proof of

$$n! = \sum_{\lambda \vdash n} (f^\lambda)^2$$

which is just the fact that the first column of the character table has unit length. A detailed discussion of these matters here would lead us too far afield. However, it is hoped that we have whetted the reader's appetite enough that they will consult the original sources.

4.11   Exercises

1. Prove the following theorem of Glaisher. The number partitions of n with no part divisible by k is equal to the number of partitions of n with each part appearing at most $k - 1$ times. Which theorem in the text is a special case of this result?

2. Let $p(n, k)$ (respectively, $p_d(n, k)$) denote the number of partitions of n into k parts (respectively, k distinct parts). Show that

$$\sum_{\substack{n \geq 0 \\ k \geq 0}} p(n, k) x^n t^k \;=\; \prod_{i \geq 1} \frac{1}{1 - x^i t}$$

$$=\; \sum_{n \geq 0} \frac{x^n t^n}{(1 - x)(1 - x^2) \cdots (1 - x^n)}$$

$$\sum_{\substack{n \geq 0 \\ k \geq 0}} p_d(n, k) x^n t^k \;=\; \prod_{i \geq 1} (1 + x^i t)$$

$$=\; \sum_{n \geq 0} \frac{x^{\binom{n+1}{2}} t^n}{(1 - x)(1 - x^2) \cdots (1 - x^n)}$$

3. Consider the plane $\mathcal{P} = \{(i, j) \mid i, j \geq 1\}$—i.e., the "shape" consisting of an infinite number of rows each of infinite length. A *plane partition of n* is an array, T, of nonnegative integers in $\mathcal{P}$ such that T is weakly decreasing in rows and columns and $\sum_{(i,j) \in \mathcal{P}} T_{i,j} = n$. For example, the plane partitions of 3 (omitting zeros) are

$$
3, \quad
\begin{matrix} 2\,1 \\[2pt] \end{matrix}, \quad
\begin{matrix} 2 \\ 1 \end{matrix}, \quad
1\,1\,1, \quad
\begin{matrix} 1\,1 \\ 1 \end{matrix}, \quad
\begin{matrix} 1 \\ 1 \\ 1 \end{matrix}
$$

a. Let

$$pp(n) = \text{ the number of plane partitions of } n$$

and show

$$\sum_{n \geq 0} pp(n) x^n = \prod_{i \geq 1} \frac{1}{(1 - x^i)^i}$$

in two ways: combinatorially and by taking a limit.

b. Define the *trace* of the array T to be $\sum_i T_{i,i}$. Let

$$pp(n, k) = \text{ the number of plane partitions of } n \text{ with trace } k$$

and show

$$\sum_{n, k \geq 0} pp(n, k) x^n t^k = \prod_{i \geq 1} \frac{1}{(1 - x^i t)^i}$$

4. a. Let τ be a rooted tree (see Chapter 3, Exercise 1). Show that the generating function for reverse τ-partitions is

$$\prod_{v \in \tau} \frac{1}{1 - x^{h_v}}$$

and use this to rederive the hook formula for rooted trees.

   b. Let λ^* be a shifted shape (see Chapter 3, Exercise 4). Show that the generating function for reverse λ^*-partitions is

$$\prod_{(i,j) \in \lambda^*} \frac{1}{1 - x^{h^*_{i,j}}}$$

and use this to rederive the hook formula for shifted tableaux.

5. The *principal specialization* of a symmetric function in the variables $\{x_1, x_2, \ldots, x_m\}$ is obtained by replacing x_i by q^i for all i.

   a. Show that the Schur function specialization $s_\lambda(q, q^2, \ldots, q^m)$ is the generating function for semistandard λ-tableaux with all entries of size at most m.

   b. Define the *content* of cell (i, j) to be

$$c_{i,j} = j - i$$

   Prove that

$$s_\lambda(q, q^2, \ldots, q^m) = q^{m(\lambda)} \prod_{(i,j) \in \lambda} \frac{1 - q^{c_{i,j}+m}}{1 - q^{h_{i,j}}}$$

   where $m(\lambda) = \sum_{i \geq 1} i\lambda_i$. Use this result to rederive Theorem 4.2.2.

6. Prove part 2 of Theorem 4.3.7.

7. Let $\Lambda^n_{\mathbf{Z}}$ denote the ring of symmetric functions of degree n with integral coefficients. An *integral basis* is a subset of $\Lambda^n_{\mathbf{Z}}$ such that every element of $\Lambda^n_{\mathbf{Z}}$ can be uniquely expressed as an integral linear combination of basis elements. As λ varies over all partitions of n, which of the following are integral bases of $\Lambda^n_{\mathbf{Z}}$: $m_\lambda, p_\lambda, e_\lambda, h_\lambda, s_\lambda$?

8. a. Let

$$e_k(n) = e_k(x_1, x_2, \ldots, x_n)$$

and similarly for the homogeneous symmetric functions. Show that we have the following recursion relations and boundary conditions. For $n \geq 1$,

$$e_k(n) = e_k(n-1) + x_n e_{k-1}(n-1)$$
$$h_k(n) = h_k(n-1) + x_n h_{k-1}(n)$$

and $e_k(0) = h_k(0) = \delta_{k,0}$, where $\delta_{k,0}$ is the Kronecker delta.

b. The *(signless) Stirling numbers of the first kind* are defined by

$$c(n, k) = \text{ the number of } \pi \in S_n \text{ with } k \text{ disjoint cycles}$$

The *Stirling numbers of the second kind* are defined by

$$S(n, k) = \text{ the number of partitions of } \{1, \ldots, n\} \text{ into } k \text{ subsets}$$

Show that we have the following recursion relations and boundary conditions. For $n \geq 1$,

$$c(n, k) = c(n - 1, k - 1) + (n - 1)c(n - 1, k)$$
$$S(n, k) = S(n - 1, k - 1) + kS(n - 1, k)$$

and $c(0, k) = S(0, k) = \delta_{k,0}$.

c. For the binomial coefficients and Stirling numbers, show that

$$\binom{n}{k} = e_k(\overbrace{1, 1, \ldots, 1}^{n})$$

$$= h_k(\overbrace{1, 1, \ldots, 1}^{n-k+1})$$

$$c(n, k) = e_{n-k}(1, 2, \ldots, n - 1)$$

$$S(n, k) = h_{n-k}(1, 2, \ldots, k)$$

9. Prove the elementary symmetric function version of the Jacobi-Trudi determinant.

10. A sequence $(a_k)_{k \geq 0} = a_0, a_1, a_2, \ldots$ of real numbers is *log concave* if

$$a_{k-1}a_{k+1} \leq a_k^2 \qquad \text{for all } k > 0$$

a. Show that if $a_k > 0$ for all k, then log concavity is equivalent to the condition

$$a_{k-1}a_{l+1} \leq a_k a_l \qquad \text{for all } l \geq k > 0$$

b. Show that the following sequences of binomial coefficients and Stirling numbers (see Exercise 8) are log concave in two ways: inductively and using results from this chapter.

$$\left(\binom{n}{k}\right)_{k \geq 0}, \qquad \left(\binom{n}{k}\right)_{n \geq 0}, \qquad (c(n, k))_{k \geq 0}, \qquad (S(n, k))_{n \geq 0}$$

c. Show that if $(x_n)_{n \geq 0}$ is a log concave sequence of positive reals, then, for fixed k, the sequences

$$(e_k(x_1, \ldots, x_n))_{n \geq 0} \quad \text{and} \quad (h_k(x_1, \ldots, x_n))_{n \geq 0}$$

are also log concave. Conclude that the following sequences are log concave for fixed n and k.

$$(c(n + j, k + j))_{j \geq 0}, \qquad (S(n + j, k + j))_{j \geq 0}$$

11. a. Any two bases $\{u_\lambda \mid \lambda \vdash n\}$ and $\{v_\lambda \mid \lambda \vdash n\}$ for Λ^n define a unique inner product by

$$\langle u_\lambda, v_\mu \rangle = \delta_{\lambda\mu}$$

and sesquilinear extension. Also, these two bases define a generating function

$$f(\mathbf{x}, \mathbf{y}) = \sum_{\lambda \vdash n} u_\lambda(\mathbf{x})\overline{v_\lambda(\mathbf{y})}$$

Prove that two pairs of bases define the same inner product if and only if they define the same generating function.

b. Show that

$$\sum_\lambda \frac{1}{z_\lambda} p_\lambda(\mathbf{x}) p_\lambda(\mathbf{y}) = \prod_{i,j} \frac{1}{1 - x_i y_j}$$

12. Prove the following. Compare with Exercise 11 in Chapter 3.

   a. Theorem 4.8.7

   b. There is a bijection $M \longleftrightarrow T$ between symmetric matrices $M \in$ Mat and semistandard tableaux such that the trace of M is the number of columns of odd length of T.

   c. We have the generating functions:

$$\sum_\lambda s_\lambda = \prod_i \frac{1}{1 - x_i} \prod_{i<j} \frac{1}{1 - x_i x_j}$$

$$\sum_{\lambda' \text{even}} s_\lambda = \prod_{i<j} \frac{1}{1 - x_i x_j}$$

13. Give a second proof of the Littlewood-Richardson rule by finding a bijection

$$(P, Q) \longleftrightarrow (T, U)$$

such that

   - P, Q, T and U are semistandard of shape μ, ν, λ, and λ/μ, respectively,

   - $\operatorname{cont} P + \operatorname{cont} Q = \operatorname{cont} T$, where compositions add componentwise, and

   - π_U is a reverse lattice permutation of type ν.

14. The *Durfee size* of a partition λ is the number of diagonal squares (i, i) in its shape. Show that $\chi^\lambda_\alpha = 0$ if the Durfee size of λ is bigger than the number of parts of α.

Bibliography

[And 76]   G. E. Andrews, *The Theory of Partitions*, Encyclopedia of Mathematics and its Applications, Vol. 2, Addison-Wesley, Reading, MA, 1976.

[Boe 70]   H. Boerner, *Representations of Groups with Special Consideration for the Needs of Modern Physics*, North-Holland, New York, NY, 1970.

[C-R 66]   C. W. Curtis and I. Reiner, *Representation Theory of Finite Groups and Associative Algebras*, Pure and Applied Science, Vol. 11, Wiley-Interscience, New York, NY, 1966.

[DKR 78]   J. Désarménien, J. P. S. Kung and G.-C. Rota, "Invariant theory, Young bitableaux and combinatorics," *Adv. in Math.* **27** (1978), 63–92.

[Dia 88]   P. Diaconis, *Group Representations in Probability and Statistics*, Institute of Mathematical Statistics, Lecture Notes-Monograph Series, Vol. 11, Hayward, CA, 1988.

[DFR 80]   P. Doubilet, J. Fox and G.-C. Rota, "The elementary theory of the symmetric group," in *Combinatorics, Representation Theory and Statistical Methods in Groups*, T. V. Narayama, R. M. Mathsen and J. G. Williams eds., Lecture Notes in Pure and Applied Mathematics, Vol. 57, Marcel Dekker, New York, NY, 1980, 31–65.

[Eul 48]   L. Euler, *Introductio in Analysin Infinitorum*, Marcum-Michaelem Bousquet, Lausannae, 1748.

[FRT 54]   J. S. Frame, G. de B. Robinson and R. M. Thrall, "The hook graphs of the symmetric group," *Canad. J. Math.* **6** (1954), 316–325.

[F-Z 82]   D. S. Franzblau and D. Zeilberger, "A bijective proof of the hook-length formula," *J. Algorithms* **3** (1982), 317–343.

[Fro 00]   G. Frobenius, "Über die Charaktere der symmetrischen Gruppe," *Preuss. Akad. Wiss. Sitz.* (1900), 516–534.

[Fro 03]   G. Frobenius, "Über die charakteristischen Einheiten der symmetrischen Gruppe," *Preuss. Akad. Wiss. Sitz.* (1903), 328–358.

[Gan 78]   E. Gansner, *Matrix Correspondences and the Enumeration of Plane Partitions*, Ph.D. thesis, M.I.T., 1978.

[G-M 81]   A. Garsia and S. Milne, "A Rogers-Ramanujan bijection," *J. Combin. Theory Ser. A* **31** (1981), 289–339.

[Ges um]   I. Gessel, "Determinants and plane partitions," unpublished manuscript.

[G-V 85]   I. Gessel and G. Viennot, "Binomial determinants, paths, and hooklength formulae," *Adv. in Math.* **58** (1985), 300–321.

[G-V ip]   I. Gessel and G. Viennot, "Determinants, paths, and plane partitions," in preparation.

[G-J 83]   I. P. Goulden and D. M. Jackson, *Combinatorial Enumeration*, Wiley-Interscience, New York, NY, 1983.

[Gre 74]   C. Greene, "An extension of Schensted's theorem," *Adv. in Math.* **14** (1974), 254–265.

[GNW 79] C. Greene, A. Nijenhuis, and H. S. Wilf, "A probabilistic proof of a formula for the number of Young tableaux of a given shape," *Adv. in Math.* **31** (1979), 104–109.

[Hai pr]   M. D. Haiman, "Dual equivalence with applications, including a conjecture of Proctor," preprint.

[Her 64]   I. N. Herstein, *Topics in Algebra*, Blaisdale, Waltham, MA, 1964.

[H-G 76]   A. P. Hillman and R. M. Grassl, "Reverse plane partitions and tableau hook numbers," *J. Combin. Theory Ser. A* **21** (1976), 216–221.

[Jac 41]   C. Jacobi, "De functionibus alternantibus earumque divisione per productum e differentiis elementorum conflatum," *J. Reine Angew. Math. (Crelle)* **22** (1841), 360–371. Also in *Mathematische Werke*, Vol. 3, Chelsea, New York, NY, 1969, 439–452.

[Jam 76]   G. D. James, "The irreducible representations of the symmetric groups," *Bull. London Math. Soc.* **8** (1976), 229–232.

[Jam 78]   G. D. James, *The Representation Theory of Symmetric Groups*, Lecture Notes in Math., Vol. 682, Springer-Verlag, New York, NY, 1978.

[J-K 81]   G. D. James and A. Kerber, *The Representation Theory of the Symmetric Group*, Encyclopedia of Mathematics and its Applications, Vol. 16, Addison-Wesley, Reading, MA, 1981.

[Kad pr]   K. Kadell, "Schützenberger's 'jeu de taquin' and plane partitions," preprint.

[Knu 70]   D. E. Knuth, "Permutations, matrices and generalized Young tableaux," *Pacific J. Math.* **34** (1970), 709–727.

[Led 77]   W. Ledermann, *Introduction to Group Characters*, Cambridge University Press, Cambridge, 1977.

[Lit 50]   D. E. Littlewood, *The Theory of Group Characters*, Oxford University Press, Oxford, 1950.

[L-R 34]   D. E. Littlewood and A. R. Richardson, "Group characters and algebra," *Philos. Trans. Roy. Soc. London Ser. A* **233** (9134), 99–142.

[Mac 79]   I. G. Macdonald, *Symmetric Functions and Hall polynomials*, Oxford University Press, Oxford, 1979.

[Mur 37]   F. D. Murnaghan, "The characters of the symmetric group," *Amer. J. Math.* **59** (1937), 739–753.

[Nak 40]   T. Nakayama, "On some modular properties of irreducible representations of a symmetric group I and II," *Jap. J. Math.* **17** (1940) 165–184, 411–423.

[Pee 75]   M. H. Peel, "Specht modules and the symmetric groups," *J. Algebra* **36** (1975), 88–97.

[Pra ta]   P. Pragacz, "Algebro-geometric applications of Schur *S*- and *Q*-polynomials," *Seminaire d'Algèbre Dubreil-Malliavin*, to appear.

[PTW 83]   G. Pólya, R. E. Tarjan, and D. R. Woods, *Notes on Introductory Combinatorics*, Birkhäuser, Boston, MA, 1983.

[Rem 82]   J. B. Remmel, "Bijective proofs of formulae for the number of standard Young tableaux," *Linear and Multilin. Alg.* **11** (1982), 45–100.

[Rob 38]   G. de B. Robinson, "On representations of the symmetric group," *Amer. J. Math.* **60** (1938), 745–760.

[Rut 68]   D. E. Rutherford, *Substitutional Analysis*, Hafner, New York, NY, 1968.

[Sag 80]   B. E. Sagan, "On selecting a random shifted Young tableau," *J. Algorithms* **1** (1980), 213–234.

[Sag 82]   B. E. Sagan, "Enumeration of partitions with hooklengths," *European J. Combin.* **3** (1982), 85–94.

[Sag 87]   B. E. Sagan, "Shifted tableaux, Schur Q-functions, and a conjecture of R. Stanley," *J. Combin. Theory Ser. A* **45** (1987), 62–103.

[S-S ta]   B. E. Sagan and R. P. Stanley, "Robinson-Schensted algorithms for skew tableaux," *J. Combin. Theory Ser. A*, to appear.

[Sch 61]   C. Schensted, "Longest increasing and decreasing subsequences," *Canad. J. Math.* **13** (1961), 179–191.

[Scu 01]   I. Schur, *Über eine Klasse von Matrizen die sich einer gegeben Matrix zuordnen lassen*, Inaugural-Dissertation, Berlin, 1901.

[Scü 63]   M. P. Schützenberger, "Quelques remarques sur une construction de Schensted," *Math. Scand.* **12** (1963), 117–128.

[Scü 76]   M. P. Schützenberger, "La correspondence de Robinson," in *Combinatoire et Représentation du Groupe Symétrique*, D. Foata ed., Lecture Notes in Math., Vol. 579, Springer-Verlag, New York, NY, 1977, 59–135.

[Sta 50]   R. A. Staal, "Star diagrams and the symmetric group," *Canad. J. Math.* **2** (1950), 79–92.

[Stn 71]   R. P. Stanley, *Ordered Structures and Partitions*, Ph.D. thesis, Harvard University, 1971.

[Stn 82]   R. P. Stanley, "Some aspects of groups acting on finite posets," *J. Combin. Theory Ser. A* **32** (1982), 132–161.

[Stn 86]   R. P. Stanley, *Enumerative Combinatorics, Vol. 1*, Wadsworth & Brooks/Cole, Pacific Grove, CA, 1986.

[S-W 85]   D. W. Stanton and D. E. White, "A Schensted algorithm for rim hook tableaux," *J. Combin. Theory Ser. A* **40** (1985), 211–247.

[Tho 74]   G. P. Thomas, *Baxter Algebras and Schur Functions*, Ph.D. thesis, University College of Wales, 1974.

[Tho 78]   G. P. Thomas, "On Schensted's construction and the multiplication of Schur-functions," *Adv. in Math.* **30** (1978), 8–32.

[Tru 64]   N. Trudi, "Intorno un determinante piu generale di quello che suol dirsi determinante delle radici di una equazione, ed alle funzioni simmetriche complete di queste radici," *Rend. Accad. Sci. Fis. Mat. Napoli* **3** (1864), 121–134. Also in *Giornale di Mat.* **2** (1864), 152–158 and 180–186.

[Vie 76]   G. Viennot, "Une forme géométrique de la correspondance
            de Robinson-Schensted," in *Combinatoire et Représentation du
            Groupe Symétrique*, D. Foata ed., Lecture Notes in Math., Vol.
            579, Springer-Verlag, New York, NY, 1977, 29–58.

[Whi 84]   A. T. White, *Groups, Graphs, and Surfaces*, North-Holland Math-
            ematics Series, Vol. 8, New York, NY, 1988.

[Whi 83]   D. E. White, "A bijection proving orthogonality of the characters
            of $\mathcal{S}_n$," *Adv. in Math.* **50** (1983), 160–186.

[Wil 90]   H. S. Wilf, *Generatingfunctionology*, Academic Press, Boston,
            MA, 1990.

[You 02]   A. Young, "On quantitative substitutional analysis II," *Proc. Lon-
            don Math. Soc. (1)* **34** (1902), 361–397.

[You 27]   A. Young, "On quantitative substitutional analysis III," *Proc.
            London Math. Soc. (2)* **28** (1927), 255–292.

[You 29]   A. Young, "On quantitative substitutional analysis IV," *Proc.
            London Math. Soc. (2)* **31** (1929), 253–272.

[Zei 84]   D. Zeilberger, "A short hook-lengths bijection inspired by the
            Greene-Nijenhuis-Wilf proof," *Discrete Math.* **51** (1984), 101–108.

Index